Solid Waste Management in Canada

This edited volume takes a comprehensive look at solid waste management across jurisdictions in Canada, including provinces, territories, municipalities, and Indigenous communities. It provides the reader with an understanding of various solid waste management approaches, policies, practices, barriers, and innovations that are being pursued and developed by jurisdictions to solve their current challenges and improve current systems in place. Solid waste management remains one of the most challenging environmental concerns in the 21st century. Understanding its complexity by bridging theory and practice is essential for current and future management, and for achieving the Sustainable Development Goals. Using a multidisciplinary approach, contributors include social scientists, engineers, economists, scientists, urban planners, and practitioners in the solid waste management field, who utilize qualitative, quantitative, and mixed methods in examining solid waste management issues in Canada. Topics include solid waste management policy and governance, community-based approaches to waste management, waste management in northern, remote, rural, and Indigenous communities, landfill management and stabilization, and innovative and emergent waste management. This book is an important resource for researchers, undergraduate and graduate students, policymakers, solid waste management professionals, government officials, and members of the public interested in solid waste management.

Anderson Assuah is an associate professor in the Aboriginal and Northern Studies program at the University College of the North (The Pas Campus), Manitoba, Canada.

Kelvin Tsun Wai Ng is Professor of Environmental Systems Engineering at the University of Regina, Canada.

Routledge Critical Discard Studies

The *Routledge Critical Discard Studies* series investigates the social, cultural, political, and economic systems of waste creation, management, and circulation. It publishes a wide range of interdisciplinary approaches drawing from the humanities and social sciences, including anthropology, environmental studies, geography, literature, politics, sociology, and science and technology studies to explore the impacts of discarded people, places, and materials. The series welcomes authored and edited volumes aimed at academics, students and researchers.

If you are interested in submitting a proposal for the series, please contact commissioning editor Grace Harrison (Grace.Harrison@tandf.co.uk) to discuss further.

Solid Waste Management in Canada: Approaches, Practices, and Experiences
Anderson Assuah and Kelvin Tsun Wai Ng

Solid Waste Management in Canada

Approaches, Practices, and Experiences

Edited by Anderson Assuah and Kelvin Tsun Wai Ng

Routledge
Taylor & Francis Group
LONDON AND NEW YORK

earthscan
from Routledge

First published 2025
by Routledge
4 Park Square, Milton Park, Abingdon, Oxon OX14 4RN

and by Routledge
605 Third Avenue, New York, NY 10158

Routledge is an imprint of the Taylor & Francis Group, an informa business

British Library Cataloguing-in-Publication Data
A catalogue record for this book is available from the British Library

ISBN: 9781032872094 (hbk)
ISBN: 9781032877228 (pbk)
ISBN: 9781003534129 (ebk)

DOI: 10.4324/9781003534129

Typeset in Times New Roman
by Newgen Publishing UK

Contents

About the Editors

Anderson Assuah is an associate professor in the Aboriginal and Northern Studies program at the University College of the North (The Pas Campus), Manitoba, Canada. His research mainly focuses on solid waste management, sustainability, environmental governance, Indigenous environmental knowledge, and climate change in northern, rural, and Indigenous communities. Dr. Assuah is the founder and chairperson of the University College of the North Sustainability Committee, which promotes Indigenous and non-Indigenous sustainability strategies in Northern Manitoba. He earned a PhD in natural resources and environmental management from the Natural Resources Institute, University of Manitoba, Canada.

Kelvin Tsun Wai Ng is Professor of Environmental Systems Engineering at the University of Regina, Canada. Kelvin's fields of interest are in sustainable solid waste management, particularly in data-driven waste policy and waste systems design. Kelvin has received awards in both teaching and research. He has reviewed over 530 journal manuscripts. Kelvin is currently serving as an associate editor for *Waste Management* (Elsevier), an editor for *Environmental Science and Pollution Research* (Springer), and an editorial board member for *Ecological Informatics* (Elsevier). He has made editorial decisions on 260+ journal manuscripts as editor or associate editor. Kelvin has organized seven international conferences.

Contributors

Gopal Achari is Professor and Head of the Department of Civil Engineering at Schulich School of Engineering, University of Calgary. He is internationally recognized for his research in water and wastewater treatment and soil remediation, particularly in advanced oxidative methods and photochemical treatments using low-energy LEDs. With over 220 publications and the supervision of 50 graduate students, Dr. Achari has received numerous awards for his research and teaching, including the APEGA Environment and Sustainability Award, AsTech Honoree, "Professor of the Year" award, and the graduate educator award. He was the former president of the Canadian Society for Civil Engineering (CSCE) and has also served as Senior VP and Chair of CSCE's Environmental Engineering Division.

Olufemi Akintunde is a postdoctoral associate at the University of Calgary. He obtained master's and doctoral degrees in chemical engineering and environmental engineering, respectively, at the University of Calgary. He also has a master's degree in chemical engineering at the University of Lagos, Nigeria. He was a recipient of Alberta Graduate Excellence Scholarship (AGES) Doctoral Research Award and has a patent with the US Patent Office. Dr. Akintunde's research interests are in water and wastewater treatment by advanced oxidation and photocatalysis processes and methane emissions mitigation by phytoremediation and bioremediation processes.

Wameed Alghazali is a research associate in the Civil and Environmental Engineering Department at Carleton University, Canada. Previously, he served as a lecturer and academic coordinator at the University of Sharjah, UAE. His research specializes in modeling the coupled processes within landfills, including thermal, hydraulic, biological and mechanical interactions, as well as reactive multi-component gas transport in porous media. He is actively involved in ongoing research focused on modeling landfill settlement and designing innovative biomethane oxidation systems aimed at enhancing landfill sustainability. Dr. Alghazali received his BSc and MSc from the University of Sharjah and his PhD from Carleton University.

Hunar Arora has a strong focus on sustainable waste management and circular economy and is currently working with the Solid Waste Resources Team at Halifax Regional Municipality. She holds a master's degree in resource and environmental management from Dalhousie University and has collaborated with organizations like Divert Nova Scotia and the Global Plastics Policy Centre. Her contributions include two published research articles and a policy brief for the INC2 of the Plastics Treaty. Dedicated to advancing impactful solutions, Hunar combines research, community engagement and policy innovation to foster sustainability in waste management.

Anderson Assuah is an associate professor in the Aboriginal and Northern Studies program at the University College of the North (The Pas Campus), Manitoba. His research mainly focuses on solid waste management, sustainability, environmental governance, Indigenous environmental knowledge, and climate change in northern, rural, and Indigenous communities. Dr. Assuah is the founder and chairperson of the University College of the North Sustainability Committee, which promotes Indigenous and non-Indigenous sustainability strategies in Northern Manitoba. He earned a PhD in natural resources and environmental management from the Natural Resources Institute, University of Manitoba, Canada.

Hadiya Bamragha was born and raised in Kenya, earned the African and Asian Khalifa Foundation scholarship to study at the United Arab Emirates University, graduating with a BSc in biology and a minor in political science. She interned at the United Nations Habitat and the International Renewable Energy Agency, sparking her interest in renewable energy and waste management. Hadiya also assessed Bonne Bay's marine ecosystem resilience and health, contributing to a MITACS-funded report. For her thesis, she collaborated with Western Regional Waste Management to study contamination issues in recycling, aiming to improve community compliance with recycling guidelines.

Ana Maria Rodrigues Costa de Castro is a PhD candidate in hydraulic engineering and sanitation at the University of São Paulo, Brazil. She is currently researching the contribution of waste pickers' integration initiatives to the just transition to a circular economy. From September 2023 to February 2024, she was a visiting researcher at the Community-Based Research Laboratory at the University of Victoria, Canada, where she studied the integration of waste pickers within the North American context. Ana holds a bachelor's degree in environmental engineering and a master's degree in science, and she has experience in research and extension projects on waste management.

Gabriella Dee holds a PhD in biology and is currently a research assistant at the School of Environmental Studies, Queen's University, Canada. She has investigated the relationship between the waste hierarchy and local waste management, as well as the importance of volunteers in reducing waste at festivals and local events. With her move from the study of life to the study of discarding objects, Dee is currently bridging the gap between natural and social sciences in the study of waste in society.

Mercy D. Fiamavle is an environmental scientist who recently earned a Master of Resource and Environmental Management from the School for Resource and Environmental Studies at Dalhousie University. With a strong passion for resource and environmental management, Fiamavle focuses on environmental assessments, pollution remediation, industrial sustainability, waste management, and occupational health and safety. Fiamavle has coauthored research on lead toxicity and is actively involved in sustainability and climate change initiatives, staying informed on emerging trends to promote innovation and minimize environmental impacts.

Jutta Gutberlet is Professor of Human Geography and affiliated with the Centre for Global Studies at the University of Victoria. She leads the Community-Based Research Laboratory. In research and practice, her work focuses on intersectional and interdisciplinary themes related to waste. She is interested in everyday livelihoods and development issues related to waste management and governance particularly in Global South countries. She brings an overlapping socioeconomic, cultural, and environmental perspective and a participatory action-oriented approach to her research. Her research involves local participants, which allows disenfranchised communities and individuals to use their agency to achieve greater social and environmental justice.

Hiroshan Hettiarachchi is currently serving as Dean of the School of Engineering at the University of Guam (UOG). He has diverse experience in geotechnical engineering, sustainable waste and wastewater management, and circular economy. Before joining UOG, he served as a professor and an academic administrator at universities in the USA and Germany and worked with the UN on global sustainability initiatives. The research and capacity development work he conducted during his stint at the UN were primarily focused on looking at waste as a resource in the context of water–food energy nexus.

J. Patrick Hettiaratchi is Emeritus Professor of Civil Engineering at the University of Calgary, Canada, specializing in solid waste management and greenhouse gas emission control. His research focuses on waste-to-energy, landfill technology, and methane emission control using biofilters. With over 200 publications, he has trained more than 50 PhD and MSc students. His accolades include the 2007 Alberta Emerald Award, the 2008 Shell/ASTech Award, and the 2013 Albert Berry Award from CSCE. Dr. Hettiaratchi has also collaborated on research projects and provided consulting services in Canada, Mexico, Chile, Ecuador, India, Thailand, and Sri Lanka.

Myra J. Hird is a professor, elected fellow of the Royal Society of Canada, and Queen's National Scholar in the School of Environmental Studies, Queen's University, Canada. Professor Hird is Director of Waste Flows, an interdisciplinary research project focused on waste as a global scientific-technical and socioethical issue. She has published 14 books and over 90 articles and book chapters on a diversity of topics relating to waste and science studies. Her 14th book, Waste: The Basics, is published by Taylor and Francis.

Poornima Jayasinghe is Assistant Professor in the teaching stream of Sustainable Systems Engineering and Civil Engineering at the Schulich School of Engineering, University of Calgary. As a chemical and process engineer specializing in environmental engineering, she is a registered professional engineer in Alberta. With extensive research experience, she has engaged in real-world projects, collaborating with private, public, and governmental organizations. In addition to her practical work, Dr. Jayasinghe teaches a variety of graduate and undergraduate courses across different programs. Her contributions to research and teaching have been recognized with several prestigious awards.

Nima Karimi is a postdoctoral fellow at the Faculty of Engineering and Applied Science, University of Regina. His research focuses on solid waste management, particularly the integration of remote sensing, machine learning, and GIS-based models to optimize waste facility locations at regional levels. Nima completed his PhD in environmental systems engineering, where his work on solid waste management systems was nominated for the University of Regina's gold medal. He has published 39 peer-reviewed articles and conference papers and collaborates with academic and industry partners to advance waste management and environmental sustainability.

Shawn Kenny is a professor in the Department of Civil and Environmental Engineering at Carleton University in Ottawa, Canada. Shawn has over 36 years of experience primarily focused on the mechanical response of linear infrastructure (e.g., culverts, energy pipelines, railways) to large-deformation ground movement events and climate change effects. He has used computational simulation tools, primarily based on finite element methods, to predict load effects on the performance and integrity of these assets. Remote sensing and field studies have been key supporting elements. Shawn received his BEng and MEng (thesis) from Memorial University and a PhD from Dalhousie University.

Calvin Lakhan is currently Director of the Circular Innovation Hub at York University's Faculty of Environmental and Urban Change, a research project devoted to advancing understanding of waste management research and policy in Canada. Calvin's current research can largely be divided into four areas: (1) evaluating the efficacy of municipal policy instruments used to promote waste diversion, (2) optimizing the recycling system to achieve increased diversion while minimizing material management costs, (3) examining the role of race and ethnicity as antecedents to recycling behavior, and (4) encouraging proenvironmental behavior among minority and marginalized groups.

Scott Cameron Lougheed was most recently a postdoctoral fellow in geography and planning at the University of Toronto. His research focuses on the intersection of sustainability, agency, and governance. Substantive interest areas are waste policy, food and food technology, and biopolitics and (bio)security.

Virginia Maclaren is Associate Professor Emerita in the Department of Geography and Planning at the University of Toronto. She investigates the social, economic and policy aspects of waste management, most recently in the Canadian context but previously in Southeast Asia as well. Her research has focused on a range of waste types and sources, including food waste, municipal waste, and multi-residential waste, and a variety of management approaches. She has served on several waste-related government advisory groups, including the board of directors for Waste Diversion Ontario and the City of Toronto's Long-Term Waste Management Strategy Stakeholder Advisory Group.

Saranga Munasinghage is a dedicated environmental professional who completed a master's in civil engineering with a specialization in environmental engineering in September 2024. Currently working as a research associate at the University of Calgary's Centre for Environmental Engineering Research & Education, he focuses on innovative research involving biocell landfill projects and application of methanotrophs to control methane emissions in the oil and gas, as well as agricultural sectors, contributing to the ongoing efforts to combat climate change and improve environmental sustainability.

Kelvin Tsun Wai Ng is Professor of Environmental Systems Engineering at the University of Regina, Canada. Kelvin's fields of interest are in sustainable solid waste management, particularly in data-driven waste policy and waste systems design. He has received awards in both teaching and research. He has reviewed over 530 journal manuscripts. He is currently serving as an associate editor for *Waste Management* (Elsevier), an editor for *Environmental Science and Pollution Research* (Springer), and an editorial board member for *Ecological Informatics* (Elsevier). Kelvin has made editorial decisions on 260+ journal manuscripts as editor or associate editor. He has organized seven international conferences.

Shailja Pant Pant holds a PhD in applied microbiology and biotechnology and has devoted her early research career to environmental biotechnology, particularly in waste-to-energy conversion processes. Her work spans fermentation technology, bioremediation, and the valorization of lignocellulosic waste for bioenergy production. Dr. Shailja is actively involved in a collaborative project between NIT Jalandhar and the University of Manitoba, aiming to develop sustainable nutrient recovery systems from wastewater. Her research is aimed at bridging the gap between environmental sustainability and technological advancements in bioconversion and resource recovery.

Dinesh Pokhrel is an environmental professional currently employed as a senior infrastructure planning engineer with the City of Calgary, where he manages projects related to landfill master plans and design of landfill components. He is also an adjunct faculty member at the University of Calgary and has assisted graduate students in their research in landfill engineering. In the past, he has been involved in the analysis, design, and modeling of landfill components and processes. He was part of the research team involved in preparing the design base report on the Calgary Biocell project in 2004.

Garrett Richards is an associate professor with the Environmental Policy Institute (EPI) at Grenfell Campus, Memorial University of Newfoundland and Labrador. While his work often involves the concrete issues of climate change, renewable energy, and waste management, his main focus is research impact. Broadly, he is interested in transdisciplinarity, partnered research, evidence-based policy, and knowledge mobilization. Relatedly, Garrett has served as graduate officer for the transdisciplinary sustainability PhD program at Grenfell Campus, coordinator of the Environmental Policy Innovation Lab at EPI, and team member with the Transdisciplinary Education Collaboration for Transformations in Sustainability (TRANSECTS) at the University of Saskatchewan.

Amy Richter is a postdoctoral fellow at the University of Regina. Within the field of solid waste management, Amy has a particular interest in integrated regional waste management systems, extended producer responsibility policy, and data-driven approaches to improve solid waste management in Canada. She was awarded the 2021 Presidents Distinguished Graduate Award for outstanding academic achievement and contributions to the university and broader community. She has published over 60 papers in a variety of top-tier journals, presented research work at international conferences and completed consultancy reports for several other projects.

Prangya Ranjan Rout is an assistant professor at NIT Jalandhar. His research focuses on bioreactor design, anaerobic digestion, and removal of emerging contaminants from wastewater. With over 100 publications, including peer-reviewed journal articles, book chapters, and conference presentations, Dr. Rout is an active contributor to international scientific discourse on waste-to-energy technologies and sustainable wastewater treatment. His research has earned recognition for addressing global environmental challenges by advancing the treatment of contaminants and promoting the sustainable reuse of waste resources. He is an active collaborator working with national and international research institutes to drive forward the development of green technologies.

Seniru Ruwanpura is a research assistant at the University of Calgary's Schulich School of Engineering and a registered engineer in training in Alberta. He completed his undergraduate degree in software engineering at Schulich School of Engineering on a full scholarship and was recognized as the best senior student of 2024 by the university's Engineering Students Society. Seniru focuses on data modeling in landfill projects and has previously engaged in software requirements engineering and responsible AI research. In 2024, he was named to Alberta's Top 30 Under 30 for his educational leadership and work in sustainability.

Christina Seidel holds a PhD in engineering management and has an extensive background in circular economy and waste reduction. She operates Sonnevera International Corp., a waste reduction consulting firm, and was Executive Director of the Recycling Council of Alberta for over 30 years, where she was involved in the waste management policy development arena and led circular economy and EPR advocacy efforts. She was a member of the National Zero Waste Council and the Canada Plastics Pact and chaired the Canadian Standards Association's "Plastics Recycling: Definitions, Reporting and Measuring" technical committee.

Delezia S. Singh is an ecotoxicology scientist and a postdoctoral research fellow at the University of Manitoba, Canada. Her current project developments focus on ecotoxicological assessment of pharmaceutical effects on crustaceans and environmental impacts of plastic bottle cap wastes. She has a multidisciplinary background spanning several fields—environmental sciences, aquatic toxicology, carcinology, and freshwater ecology. Her research expertise involves the measurement of contaminants in different environmental compartments to better predict their adverse risks to organisms, designing assessments for biomarker identification, endpoints and mechanistic toxicity testing for different organic (e.g., pesticides) and inorganic (e.g., heavy metals) pollutants.

Anica Tasnim is an MASc student in environmental systems engineering at the University of Regina. Her research focuses on waste management, with a particular interest in resource recovery from waste and landfill management. Anica has published two papers on waste management in Canadian jurisdictions in top-tier journals and has contributed to five additional research projects as a research assistant during her master's journey. Committed to advancing her expertise in the field, she continues to collaborate with academic team members on projects related to waste management, aiming to strengthen her research experience and impact in environmental sustainability.

Pradeep Thapa holds an MSc degree in civil engineering from the University of Manitoba. Currently he is duty manager of engineering work at Delta Hotels by Marriott.

Thoreau Rory Tooke is the Manager of Sustainability and Asset Management at the City of Victoria, British Columbia. He established and leads the solid waste engineering and planning function of the city with responsibilities for zero waste and circular economy policies and programs, planning and design of municipal solid waste services and corporate sustainability initiatives. Rory is the current Vice Chair of the Capital Region's Solid Waste Advisory Committee. He holds a master's and PhD in natural resources management from the University of British Columbia.

Paul J. Van Geel is a professor in the Department of Civil and Environmental Engineering at Carleton University in Ottawa, Canada. His expertise is primarily focused on the transport and fate of contaminants in the subsurface and issues related to waste management. His research related to waste management has focused on anaerobic digestion of solid waste and landfill gas to energy; optimization of waste stabilization in landfills operated in northern climates; compost-based landfill biocovers; nuclear waste disposal and waste-to-energy. Dr. Van Geel received his BASc and PhD in civil engineering from the University of Waterloo.

Tony R. Walker is a professor in the School for Resource and Environmental Studies, Dalhousie University. His research efforts currently focus on a wide range of biophysical sciences related to resource and environmental management topics, including monitoring, management, and mitigation of plastic and microplastic pollution; industrial contaminants; remediation and reclamation of contaminated sites; ecological risk assessment of aquatic and terrestrial ecosystems; air pollution impacts on ecological and human health; environmental effects monitoring; environmental impact assessment; and management of natural resources. According to Scopus, he has coauthored papers with 793 different researchers making a high global impact.

Qiuyan Yuan is an associate professor at the University of Manitoba and is also Associate Head of Undergraduate Programs in the Department of Civil Engineering. She earned her PhD in civil engineering from the University of Manitoba in 2012, following a BSc in food science engineering from Beijing Agricultural University. Dr. Yuan is widely recognized for her contributions to environmental engineering. Her pioneering work in nutrient recovery focuses on innovative biological and chemical processes for phosphorus removal and recovery, aiming to develop sustainable technologies that reduce greenhouse gas emissions and environmental burdens. She has led several interdisciplinary research projects.

Acknowledgments

There are many who have contributed to the successful completion of this book, and we would like to publicly acknowledge them.

We foremost acknowledge all the contributors of this edited volume for their hard work and tireless efforts in working with us to complete this book. Without their decision to share the knowledge learned from communities and information gathered from various sources, this book could not have been written. We are grateful to them for making this book a valuable resource for solid waste management in Canada. We adore their commitment and dedication.

We also acknowledge all research participants who took time out of their schedules to be interviewed, surveyed, as well as provide invaluable documents and information to the contributors. We thank them for sharing their insights, knowledges, and experiences. Similarly, we are thankful to the communities—Indigenous and non-Indigenous—for allowing research to be conducted in their various territories and jurisdictions.

To the many municipalities, organizations, policymakers, research agencies, and institutions that provided financial support for conducting research and other kinds of assistance, thank you. We also thank all the dedicated editorial staff at Routledge.

We deeply acknowledge our families for the support throughout this long journey, which sometimes meant that we sacrificed family time.

Anderson would like to thank his wife, Shelby, for the continued prayers, support, and encouragement throughout the process. Your kind words and commitment toward my academic career is profound, and I am grateful. To my young children, Zurielle and Avniel, who may not fully grasp what was at stake but sometimes stayed with me throughout the night, I appreciate your time, laughter, and care. Your words, "Daddy, are you okay?" to check up on me while working were a source of encouragement to me. I am thankful to my friends, colleagues, and mentors, who continue to inspire me to be the best version of myself. To Mr. Thaddeus Assuah and Mrs. Gladys Assuah, my late parents, your words of wisdom continue to reverberate and guide me. I am proud to have you as my parents. Above all, I am grateful to the Almighty God for leading my path in life.

Kelvin wishes to first express his appreciation to the dedicated work by his brilliant coeditor Anderson Assuah. This has been a wonderful collaboration. I thank my current and former students, friends, colleagues, and mentors at the University of Regina, whose understanding and patience have been essential to the book's completion. I thank my wife Jennifer who has been with me since the very beginning of my academic career. She is a fountain of inspiration in my professional work and daily life. Finally, and most importantly, I would like to thank my parents for their unconditional love and encouragements throughout my life.

Introduction

Anderson Assuah and Kelvin Tsun Wai Ng

Managing solid waste remains one of the most challenging environmental concerns in the 21st century, because its generation continues to increase globally while management approaches are not commensurate with the amounts being produced. Without careful management and serious action toward its management, the municipal solid waste generation of about 2.1 billion tonnes in 2020, for instance, will increase to approximately 3.2 billion tonnes in 2040 and 3.8 billion tonnes by 2050 (United Nations Environment Programme, 2024). This global outlook is not different from Canada. In fact, Canadians generate more solid waste per capita than most industrialized nations but recycle less. While there have been numerous efforts—policies, strategies, and approaches—to deal with the situation in the country, generation continues to increase (Environment and Climate Change Canada, 2024). Landfilling or permanent disposal of waste remains the main management approach in Canada (Government of Canada, 2024), hindering the country from achieving sustainable solid waste management.

Strategies and approaches to manage solid waste vary among provinces and municipalities in Canada, and some regions in the country need more support. For instance, Indigenous communities face many difficulties and roadblocks because of factors such as huge deficits in waste management infrastructure (e.g., landfills and waste transfer stations), historical underfunding for solid waste management by Indigenous Services Canada, lack of diversion programs, and lack of waste management planning and policies. The long distances away from major waste facilities also present a major challenge to many northern communities that may be pursuing initiatives to deal with waste. This is worsened by the very cold and long winters that make it challenging to operate landfills, for example, because of frosty soils. There is also the challenge of participation in solid waste management programs by community residents and a lack of understanding of sustainable waste management, even though every human generates waste as part of their very living existence.

The above wide-ranging concerns underscore the importance of understanding how waste is being generated, treated, diverted, and managed in Canada. Understanding the complexity and complicated nature of current waste management approaches and methods is essential to providing recommendations and

DOI: 10.4324/9781003534129-1

solutions to provinces and municipalities and forecasting future management options and approaches to deal with challenges and bridge any gaps so that waste management goals can be met. Doing so calls for the (re)examination of waste policies in Canadian jurisdictions, in addition to how municipalities—cities, towns, and villages—are managing the waste they generate.

This is the reason and purpose for this edited volume. The contributors, made up of social scientists, engineers, economists, scientists, urban planners, practitioners, and public officials in the solid waste management field, delve deeper into the efforts, experiences, and programs by provinces, municipalities, rural, urban, northern, Indigenous communities as well as organizations working to improve practices and systems. This book, which is the first of its kind in Canada, provides the reader with an understanding of various solid waste management policies, approaches, practices, barriers, and innovations that are being pursued and developed by jurisdictions to manage different waste streams such as electronics, plastics, paper, and industrial, commercial, and institutional waste.

Unlike similar books in solid waste management, a multidisciplinary approach is adopted in the creation of this book. Qualitative, quantitative, and mixed methods have been utilized by contributors to examine programs, initiatives, innovations, and emergent waste across the country. This makes the book ideal for researchers, teachers, graduate, and undergraduate students, as it provides varied research methods and designs to examine solid waste management in a large, industrialized nation in addition to the vital information provided on solid waste. The book can be used as a textbook in a variety of courses, such as environmental sustainability, environmental planning, environmental engineering, and discard studies. Additionally, solid waste management professionals, government officials, and policymakers will benefit from using this book, because it provides them with information about solid waste practices, examples, and strategies being utilized in different geographic regions in Canada that will guide their work. Given that solid waste management is a universal challenge that most countries across the world face, the experiences, lessons, insights, and contents of this book are transferable to the context of other countries.

This book contains 17 chapters that are organized into five parts, in the following order: "Waste management policy and governance"; "Waste management in northern, remote, rural and Indigenous communities"; "Community-based waste management approaches"; "Landfills and innovative waste management"; and "Emerging waste management."

Part 1: Waste management policy and governance

In Chapter 1, Dee and Hird examine how municipalities can achieve the higher tiers on the waste management hierarchy (e.g., reuse and repair) by examining the waste management system in the City of Kingston, Ontario. The authors analyze the challenges associated with moving up the ladder and the adjustments required by municipalities. Similarly, Tooke and Siedel, in Chapter 2, draw on the City of

Victoria's Zero Waste Victoria plan to provide insight and direction to municipalities to prioritize waste prevention and reuse to support their zero waste goals when planning for municipal solid waste. The industrial, commercial, and institutional sector in Nova Scotia significantly contributes to the province's waste generation, and the lack of compliance by stakeholders in this sector could hinder the province's policy to reduce its waste. Arora and Walker analyze the challenges in the sector and the opportunities that exist to improve compliance in Chapter 3. An important vehicle to operating in the higher tiers of waste management, reducing waste, and/or ensuring compliance is how to effectively communicate to a variety of stakeholders. Lakhan covers inclusive communication with diverse audiences in promoting sustainable waste management practices in Chapter 4.

Part 2: Waste management in northern, remote, rural and Indigenous communities

In Chapter 5, Richter et al. examine accessibility challenges to extended producer responsibility programs in northern Saskatchewan and highlight sparse population in that region of the province as a major contributing factor. Through extensive analysis of data, the authors develop a map that shows locations that would benefit from siting recycling depots in this region to encourage effective and efficient management of waste. Generally, there is limited waste management research in northern, remote, and rural communities in Canada, which makes it difficult for researchers and policymakers alike to understand local-level challenges and proffer solutions. To bridge this research and policy gap, Assuah, in Chapter 6, investigates municipal solid waste management among Northern Manitoba (north of 53rd parallel) municipalities to spotlight key factors that need to be given attention. In Chapter 7, Assuah also examines operational challenges that First Nations—who face huge challenges with managing their waste—are confronted with and the strategies waste managers and operators are using to overcome them and to be effective.

Part 3: Community-based waste management approaches

The activities and importance of waste pickers or binners to the waste management value chain are examined in Chapter 8 by Castro and Gutberlet, who focus their analysis on two grassroots organizations—the Binners' Project in Vancouver and the Diverters Foundation in Victoria—in British Columbia. The authors fill an important gap in the literature regarding waste pickers in high-income countries like Canada. Lougheed and Maclaren, in Chapter 9, examine *recycle right*, which is a campaign that urges residents in the province of Ontario to reduce recycling contamination. In Chapter 10, Bamragha and Richards surveyed residents of Western Newfoundland to understand attitudes and behaviors regarding non-compliance with the province's recycling guidelines and suggest ways to increase compliance.

Part 4: Landfills and innovative waste management

Landfill is the ultimate endpoint for all solid waste management systems. The cold winter months in Canada can result in the freezing of landfills. To understand the stabilization process of waste in landfills, Alghazali et al. discuss the outcome of a 12-year research project at the Sainte Sophie Landfill in Québec, Canada, in Chapter 11. In Chapter 12, Jayasinghe et al. also discuss the Landfill Biocell concept in Calgary, Alberta, which Canadian researchers developed to address challenges associated with sustainably operating waste cells in cold and semi-arid conditions. In Chapter 13, Thapa et al. review the potential of recovering phosphorus from food waste in Canada instead of disposing it in landfills and increasing its greenhouse gas generation. They discuss the most and least favorable food waste treatment methods from three food waste treatment scenarios. Given its large spatial footprint, landfill closure remains a key barrier in sustainable waste management. In Chapter 14, Tasnim et al. assess the potential of converting closed landfills in Saskatchewan into solar energy sites by examining topographical factors and solar radiation utilizing ArcGIS for spatial analysis and PVWatts for solar radiation estimation.

Part 5: Emerging waste management

Singh analyses emerging municipal solid waste management streams in Canada and provides insight into their potential environmental impacts if not properly planned for and managed, in Chapter 15. In Chapter 16, Fiamavle and Walker focus their analysis of emerging waste on the end-of-life of wind turbine blades in Nova Scotia and provide solutions for managing it sustainably. In the final chapter of the book, Chapter 17, Karimi et al. explore the integration of geographical information system techniques and remote sensing indices to enhance municipal solid waste management systems in Canada, including applications in waste facilities siting, landfill fugitive gas hotspots detection, and waste management regionalization.

Solid waste management was traditionally seen as an engineering function. It was soon realized, however, that problems and challenges associated with waste are complex and that no single discipline can solve our modern waste issues. This multidisciplinary approach to managing waste is a key theme of this book. This book also integrates state-of-the-art waste research from rural, remote, Indigenous, and northern regions that are not mainstreamed in waste management conversations despite their unique challenges, thereby directly addressing a missing piece of sustainable waste management puzzle in North America. The use of data-driven techniques and the development of evidence-based waste policies are increasingly popular in the design of a sustainable waste management system. Some of the key barriers to achieving this are inconsistent waste definitions and reporting periods, incomplete waste dataset, and lack of a regionalized approach. It is noteworthy that there is probably no universal prescription to our modern-day waste issues; therefore, technical merits, social and cultural values, political environment, and local legal frameworks must be thoroughly considered on a case-by-case basis. This book is a valuable resource in planning, designing, and managing a sustainable and resilient waste management system in Canada and beyond.

References

Environment and Climate Change Canada (2024). *Canadian environmental sustainability indicators: Solid waste diversion and disposal.* www.canada.ca/content/dam/eccc/documents/pdf/cesindicators/solid-waste/2024-nov/solid-waste-diversion-disposal-en.pdf

Government of Canada. (2024, June 21). *Municipal solid waste management in Canada.* www.canada.ca/en/environment-climate-change/services/managing-reducing-waste/municipal-solid/environment.html.

United Nations Environment Programme. (2024). *Global waste management outlook 2024: Beyond an age of waste: Turning rubbish into a resource.* https://wedocs.unep.org/20.500.11822/44939.

Part 1

Waste management policy and governance

1 Moving up the waste hierarchy

A Canadian case study

Gabriella Dee and Myra J. Hird

Introduction

A significant contributor to climate change and pollution, waste is a global environmental and human health crisis (United Nations Environment Programme, 2012). While the United Nations Sustainable Development Goal (SDG) 6—clean water and sanitation—most directly concerns waste issues, waste is integral to all SDGs, including 12 and 13; responsible production and consumption, and climate action, respectively. Globally, humans generate an estimated 2.1 billion tons of municipal solid waste (MSW) consisting of household garbage, recyclable and compostable materials, as well as waste material from businesses, institutions and construction and demolition sites, each year (United Nations Environment Programme, 2024). Cities are struggling to cope with mounting MSW amid increasing urbanization and "significant investment is needed in the development and maintenance of waste management infrastructure, especially in low- to middle-income countries" (United Nations 2022, p. 51). Although high-income countries account for only 17 percent of the world's population, they generate a disproportionate 34 percent of the world's waste (Kaza et al., 2018). A prime example is Canada, reported to have the highest MSW generation per capita among OECD member countries (Kawai & Tasaki, 2016), and therefore a significant contributor to multiple planetary crises.

The Waste Hierarchy (WH—also known as Lansink's Ladder) is typically illustrated as a multi-tier pyramid with the categories prevent, reuse, recycle, incineration, and disposal from top to bottom and organizes waste's management responses from most to least preferable for the environment and human health (Lansink, 2017) (Figure 1.1).

Canada's federal government adopts a WH that prioritizes reduction followed by reuse, refurbishing, recycling, and recovery, with landfilling as the least preferred option (Figure 1.1). The Canadian WH is illustrated as an upside-down triangle with reduction as the highest, widest tier, and landfill as the bottom, narrowest tier to illustrate that reduction should be "the first priority … both at the manufacturing level and by consumers and institutions" (Government of Canada, 2021). Yet, in Canada, entrenched and regulated modes of managing waste heavily rely on disposal (landfill, the bottom most tier), incineration (in the form of waste-to-energy plants, one tier up from landfill) and diversion (recycling /composting, a

DOI: 10.4324/9781003534129-3

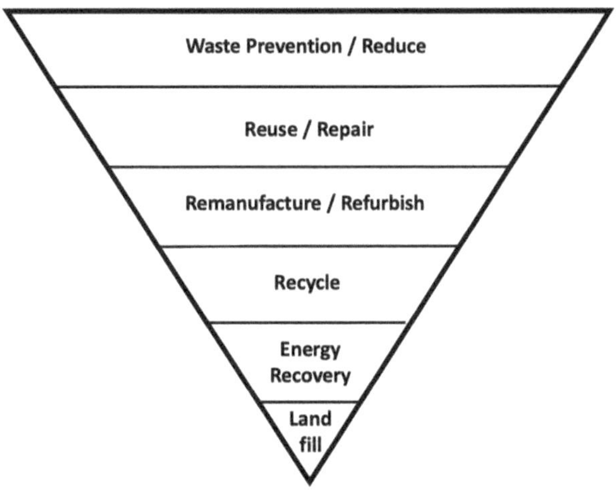

Figure 1.1 The Waste Hierarchy.

Source: Adapted from the Government of Canada (2021). Areas positively correlate to preference, with the widest bars being the most preferred strategy.

middle tier; Bulkeley et al., 2007; Pollans, 2017; Kaza et al., 2018; Hird 2021). The lowest tiers (disposal, waste-to-energy and recycling) serve the privatized waste management system. For example, private landfills accounted for 53 percent of landfill capacity in Ontario in 2021, up from 36 percent in 2017 (Ontario Waste Management Association, 2021). Greater amounts of waste translate into greater profits and more economic growth within municipalities. For instance, a Province of Ontario (2021) webpage features one of the largest water bottling companies in North America: Ice River Springs. The company employs 400 people throughout its five Ontario facilities, purchases approximately 85 percent of its polyethylene terephthalate (PET) from Ontario's recycling program to make their own 100 percent post-consumer PET recycled bottles, "and can take even more" (Province of Ontario, 2021).This illustrates the positive feedback between waste generation and economic growth.

This chapter presents the City of Kingston, Ontario, as a case study of the municipal government's efforts to reduce MSW. Kingston's waste reduction efforts reveal a tension common to municipalities across Canada, between efforts that reduce waste but do not serve Kingston's consumer culture and private enterprise, and a firmly entrenched waste management system that supports the WH's lowest tiers.

Waste management at municipal, provincial and federal levels

Globally, local governments are mainly responsible for legislating and managing MSW and liquid waste (Kaza et al., 2018; Sawell et al., 1996). In Canada, the federal government manages MSW at federally owned facilities through the National Parks Act and the Indian Act, as well as legislates the transportation of MSW both regionally and internationally while provinces and territories are responsible for the laws governing the siting, licensing, and monitoring of waste disposal facilities (Sawell et al., 1996). For this reason, regulations vary between provinces and territories, and are based on regional and political factors such as geography, population density, the political party in power, economic ties with neighboring regions, and so on (Sawell et al., 1996).

Canada primarily manages its waste through disposal. According to Environment and Climate Change Canada (ECCC, 2022) 72.5 percent of solid waste generated in 2020 in Canada was sent for disposal. Approximately 97 percent of waste meant for disposal is landfilled and 3 percent are incinerated (Government of Canada, 2024a). When normalized for population growth, ECCC (2022) reports that between 2006 and 2018 the amount of residential MSW disposal per person decreased by 2 percent (Figure 1.2). Diversion through recycling has led to a 15 percent increase in recycled and composted waste. When data for disposed and diverted waste are combined, it demonstrates that Canadians increased their overall residential MSW by 2.7 percent. It should be noted that although the Global Waste Management Outlook (United Nations Environment Programme, 2024) states that North America has 0 percent uncontrolled waste, Canada practices open dumping in Nunavut (Hird, 2021, 2022).

Ontario primarily manages its waste through disposal (landfilling) and diversion (recycling and composting). When normalized for population size, disposed residential MSW per capita in Ontario increased by 3 percent from 2006 to 2018, while diverted residential MSW increased by 16 percent (ECCC, 2022; Figure 1.2). When disposed and diverted waste are combined, Ontario residents increased their overall residential MSW by 7.2 percent—a greater increase than the average Canadian resident.

According to Statistics Canada (2021b), Ontario municipalities spent $1.323 billion on waste management in 2018, up from $592 million in 2002. About 40 percent of this expenditure consists of waste collection and transportation, and 11 percent is spent on disposal (primary landfills facilities; Statistics Canada, 2021b). Increasing waste and recycling generation is leading to significant increases in transportation (Hird & Kuyvenhoven, 2020). Where waste was once transported to small, local landfills, waste is increasingly transported to mega commercial landfills, with 60 percent of Ontario's disposal capacity located in only seven landfills throughout the province (Ontario Waste Management Association, 2021). This said, waste disposal often involves shorter transportation distances, and thus fewer carbon emissions, than recycling, as the multi-stage transfer stations and recycling facilities are often significantly further away (Hird & Kuyvenhoven, 2020). This often means that while recycling appears higher up on the WH than disposal, it incurs higher environmental and economic costs (Gille, 2007; MacBride, 2013; Zink & Geyer, 2019, Hird & Kuywenhoven, 2020).

Currently, the only noteworthy legislative initiative aimed at the WH's higher tiers is the recent Right to Repair. The federal government's 2023 budget included the aim of working toward a "targeted framework for home appliances and

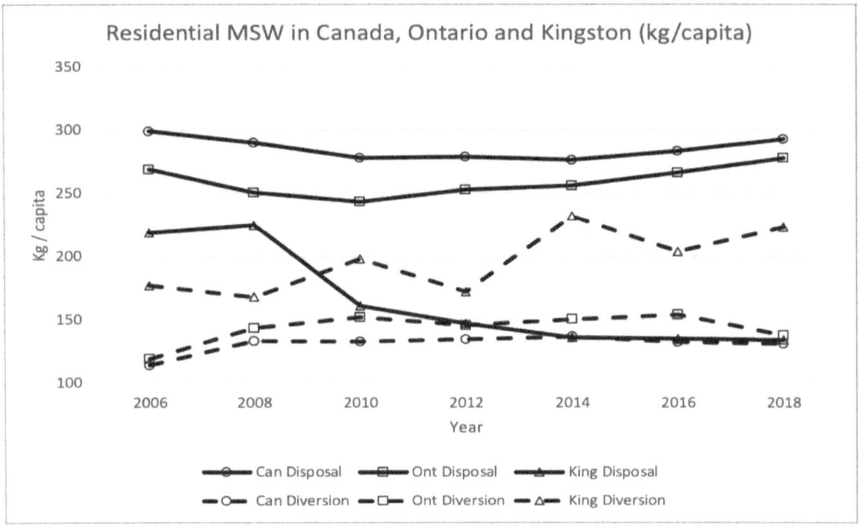

Figure 1.2 Residential MSW in kilogram per capita for residents of Canada.

Note: Residential MSW in kilogram per capita for residents of Canada (circles), Ontario (squares), and Kingston (triangles), separated by disposal (landfilling, solid lines) and diversion (recycling and composting, dotted lines).

Sources: Statistics Canada (2023, 2022a, 2022b, 2021a), and Resource Productivity and Recovery Authority (2024).

electronics" Right to Repair (Government of Canada, Budget 2023, p. 37). The federal government is in the process of amending the Copyright Act (to enable consumers to repair their devices where they choose) as well as the Competition Act (to prevent manufacturers from refusing to provide parts, tools or software needed to fix devices; Government of Canada, 2024b). The 2024 budget announced that the government would launch consultations in June to develop the Right to Repair framework, and calls on provinces and territories to amend their contract laws to support the Right to Repair legislation (Government of Canada, 2024c). Ontario's current government voted down a Right to Repair bill in May of 2019. More recently, Liberal MPP Michel Coteau's Bill 72: An Act to Amend the Consumer Protection Act 2002 sought to require brand holders such as cell phone manufacturers to provide and charge consumers, at limited rates, with the tools needed to repair their electronic products (Wilkinson, 2020).

Methodology

The case of Kingston, Ontario

Kingston is a medium sized city on the northeastern shore of Lake Ontario. According to the 2021 census of the metropolitan area, Kingston has a population of approximately 172,546 people living in a 1,919 square kilometer space (population density of 89.9 people per square kilometer) and with a median after tax income of CA $73,500 (Statistics Canada, 2022c). The population is made up of 64.2 percent working age people (15–64 years), 14.4 percent children (0–14 years) and 21.4 percent people over 65 years of age.

The present study draws from an original study of Kingston's waste management from 2005 to 2014 (Hird et al., 2014). This study employed a survey, key informant interviews, participant observation, and archival research to investigate Kingston's WM legislation, policies, practices and claims-making. The results of the studies (Lougheed et al., 2016; Hird et al., 2014) revealed that the City emphasizes individual behavioral change as the primary means of resolving its mounting waste issues that ultimately obfuscates potentially more profound systemic or upstream changes that would reduce waste.

Data collection methods

This chapter focuses on changes to Kingston's MSW management system from 2014 to 2024 and involved a comprehensive literature review and archival research to examine City Council web-based documents, such as the budgets and integrated waste management plans. Information from Statistics Canada was gathered and used to develop graphs for Canada, and Ontario. City of Kingston waste statistics were gathered through the Resource Productivity and Recovery Authority website. Coauthor Gabriella Dee employed participant observation and gathered observational data from waste processing center tours and volunteered at events such as the Memorial Centre Farmers Market, and the inaugural Sustainability Hub sponsored by the city of Kingston and Queen's University. She also conducted two key informant interviews with a member of the Solid Waste Department of the City of Kingston. The interviews were semi-structured and lasted for about an hour. The study was approved by Queen's University's General Research Ethics Board.

Kingston's waste management system

Including targets in legislation has become a way that local governments attempt to build MSW reduction into city planning (Bulkeley et al., 2007). In 2002, Kingston set a target of a 38 percent diversion rate in order to "protect the health, safety and natural environment of our citizens through fiscally responsible … practices that encourage waste reduction and recycling and that promote economic prosperity" (Hird et al., 2014, p. 449). In 2006, the City set a new target of 65 percent diversion of household waste by 2012 (Dee, 2023), and introduced organics collection in some areas, which remains integral to Kingston's Waste Strategies (City of Kingston, 2024). Having not achieved its 2012 target, in 2015 the City reset its target to 60 percent diversion by 2018 and 65 percent diversion by 2025 (Dee, 2023). According to the City's Integrated Waste Management Plan, its top waste management strategy is the "reduction of waste through behavioral/technological change" (Dee, 2023, p. 15).

However, reduction appears to apply only to waste designated for landfill, and not to total waste generated. The City of Kingston Waste Strategies webpage states that 15 percent of household waste is divertible recyclables (such as plastics, glass, and paper) and 35 percent divertible organics (according to curbside audits). So, if Kingstonians simply put their waste into the "right box or bin, Kingston would have a waste diversion rate of 76 percent!" (City of Kingston, 2024). This singular focus on increasing diversion rates explains why the current three proposed changes for the city's waste management are: a) weekly clear bag waste collection; b) biweekly clear bag waste collection, and; c) biweekly waste collection and increased bag tag fees (City of Kingston, 2024, n.d.).

The purpose of the clear bags is to aid waste collectors in refusing collection if they can see materials in the bags that could be diverted, while the change to biweekly collection would allow for the collection of yard waste on alternate weeks, leading to cost savings in processing organic waste. Increasing bag tag fees (the fees that residents pay to put more than one weekly garbage bag out to the curb) are a disincentive to residents to produce garbage destined for landfill while increasing the waste they divert. Nowhere in these proposals is there any mention of strategies to aid residents in reducing the overall amount of waste generated. The City's focus on reducing disposal waste did result in a 39 percent reduction in per capita residential MSW destined for landfill from 2006 to 2018 (Resource Productivity and Recovery Authority, 2024; see Figure 1.2), an impressive statistic when compared to the increase in disposal waste for the average resident of Ontario (3 percent) or the slight decrease for the average Canadian (2 percent; Figure 1.2). Kingston residents also reduced their per capita total generated residential MSW by 15.9 percent between 2006 (397 kg/capita) and 2022 (334 kg/capita; Resource Productivity and Recovery Authority, 2024). However, there remains no legislation targeting the top WH tiers, and the targets set in 2006 for city residents to reach 65 percent household waste diversion had yet to be met in 2023.

Landfilling

The City of Kingston contracts private waste management companies to collect and dispose of its MSW (member of the Solid Waste Department of the City of Kingston, personal communication, September 17, 2021). In a purported effort to reduce MSW generation, the City instituted a one free-bag policy in 2021, with each subsequent bag costing households $2.00 (Hird, 2021). Most of the residential MSW collected in Kingston is transported by large capacity diesel trucks to three Ontario landfills: Twin Creeks Landfill in Alvinston (490 km west), Eastern Ontario Waste Handling Facility in Moose Creek (205 km northeast), and Cornwall Landfill (175 km east). In 2020 alone, transporting MSW to landfill generated 165 tons of CO_2 eq emissions.[1] Industrial, commercial and institutional (ICI) waste is collected by a small number of companies, including Environmental 360 Solutions, Waste Connections, Waste Management, and the City of Kingston through Waste Management, Inc.

Recycling

As previously stated, the City of Kingston is focused on diverting 65 percent of its residential waste from landfill by 2025, through increased recycling (Dee, 2023). Despite the 24 percent increase in per capita diversion of residential MSW between 2006 and 2018 (Figure 1.2), the diversion rate hovered just under 60 percent for the last decade (Resource Productivity and Recovery Authority, 2024). These data indicate that diversion is the major mode of waste governance in Kingston.

The province of Ontario and the City of Kingston classify recyclables (including organic material) as diverted waste (Resource Productivity and Recovery Authority, 2024). The weekly collection of glass, plastics, metals, paper and cardboard began in the early 1990s. These materials are collected by the City and contracted private companies, and brought to the Kingston Area Recycling Centre (KARC). Although a recycling center by name, KARC is a materials recovery facility only, where recyclables are sorted, baled and sold to companies or brokers[2] that are often a significant distance from Kingston, as Table 1.1 details (see also Hird & Kuyvenhoven, 2020), and in some cases are only the first stop in a multi-stage recycling process. Since 2009, the City contracted with private company Tomlinson Organics (20 km from Kingston) to aerobically compost organic waste. The City currently collects organic waste weekly, bagged or binned yard waste twice a year, and brush twice a year. Unlike garbage bags—where the first one is free and every subsequent bag must be paid for—the City does not restrict recycled waste.

Recycling is on the fourth tier of the Government of Canada WH (Government of Canada, 2021; Figure 1.1) because of its significant energy and resource requirements, and because it creates its own waste (Gille, 2007; MacBride, 2013; Zink & Geyer, 2019). As such, waste researchers increasingly argue that recycling prioritizes "state program building and associated economic activity, not material progress in reducing waste" (MacBride, 2013, p. 83). Recent exposés lay bare the fossil fuel industry's clear knowledge that the majority of plastics (which make up the bulk of MSW recycling) cannot materially be recycled, and that those few plastics types that can be recycled are not a solution to plastics waste (Allen et al., 2024; Greenpeace, n.d.).

Of the materials collected within the municipality of Kingston, many are recycled into products that are not themselves recyclable (Table 1.1). In addition, as Table 1.1 shows, transporting materials to a recycling center often produces more carbon emissions because recycling centers are further away than landfills. For instance, polystyrene foam is transported over 1,000 km to Indianapolis. Moreover, KARC sends approximately 13 percent of the material it receives to landfill because those materials are either not recyclable or are too contaminated to be accepted (Member of Solid Waste Services Department, written communication, January 7, 2022). As a result, the City's push to encourage diversion creates what many in the industry refer to as "wish-cycling," whereby residents put materials into their diversion bins in either the hopes that those materials will be recycled, or the preference to not fill (and pay for) garbage bags with what they know to actually be garbage (Altman, 2021).

Table 1.1 Major recycling waste streams: Quantities, destinations and end uses

Material group	*Quantity 2020 (tons)*[a]	*Percent of total recycled material*[b]	*First destination(s)*	*End uses for recycled materials*
#1 PET plastic	501 t	6.8%	Quebec, Northeastern United States	Fleeces, polyesters
#2 HDPE plastic bottles	148 t	2.0%	Southwestern Ontario, Quebec, Northeastern United States	HDPE piping and stiffeners for luggage
#3-#7 mixed plastics	229 t	3.1%	Southwestern Ontario	Plastic pellets for manufacturing or resorted for further processing
Aluminum	213 t	2.9%	United States	New beverage containers
Steel	299	4.1%	Southwestern Ontario	New steel containers
Paper	3,142 t	42.7%	Ontario (Napanee), (Quebec infrequently)	Recycled paper products, paper towels, cardboard
Corrugated cardboard, boxboard	2,006 t	27.3%	Ontario, Sold to broker with unknown/ variable final destination	Boxboard, newspaper, paper towels, molded pulp
Glass	639 t	8.7%	Quebec, Ontario	Aggregates, industrial abrasives, blasting material
Polystyrene foam	6 t	0.08%	Quebec and Indianapolis	Receipt cores, grading materials

Note:

[a] Information for products, destinations and tonnage provided by Member of Solid Waste Services Department, written communication, January 7, 2022.

[b] Based on a total of 7,352 tons including 82 tons of stretchy plastic bags and 87 tons of polycoat.

The WH's higher tiers

In 2009, the Canadian Council of Ministers of the Environment (CCME) published a Canada-Wide Action Plan for Extended Producer Responsibility (EPR). Objectives of this plan include shifting the expenses associated with product end-of-life management from taxpayers to producers and consumers, reducing generated and landfilled waste, increasing diverted waste, and contributing to the harmonization of programs across the country (CCME, 2009). The plan further states that

"EPR programs are consistent with the 4R waste management hierarchy: Reduce, including reduction in toxicity and redesign of products for improved reusability or recyclability, Reuse, Recycle, Recovery of materials and/or energy" (CCME, 2009, p. 16). The EPR objective with the greatest potential to reduce waste is Design for Environment, which contributes "to natural resource and energy conservation, biodiversity preservation, source reduction, waste minimization, and pollution prevention" (CCME, 2009, p. 16). Each province is responsible for its own EPR program. Ontario began transitioning to a new EPR framework in June of 2023 in the City of Toronto, and will continue to bring municipalities on board until all municipalities adapt to this new framework by 2026. EPR is scheduled to reach the City of Kingston on July 1, 2025 (Blue Box Transition Schedule, 2024), so it is too early to tell what repercussions may ensue, and whether EPR will actually lead to any reduction in waste generation.

There are indications that some institutions, small businesses and grassroots organizations are leading waste reduction initiatives that focus on the WH's higher tiers. In 2023, the City hosted its first ever Sustainability Hub (in conjunction with Queen's University), which featured individuals from Kingston's new Repair Café. The Repair Café opened in 2017 with a volunteer staff (in partnership with the Kingston Frontenac Public Library) to teach community members how to repair their broken items. For two afternoons in March 2023, several sustainability groups from Queen's University, Repair Café volunteers and members of Cycle Kingston gathered to help community members repair furniture, small appliances, clothing and bicycles.

Several businesses and organizations within the city of Kingston promote reuse as well as waste reduction, though it should be noted that their products/services are more expensive than disposable alternatives. Three refilleries that sell products such as cleaning supplies and toiletries in reusable containers opened downtown in the last few years. Limestone Creamery, an organic farm, supplies milk in returnable/reusable glass bottles, and Mumma Duck's Diaper Services provides laundering services for cloth diaper customers. The Memorial Centre Farmers Market, the city's only year-round market, started a reusable food service-ware initiative in 2023, providing vendors who serve prepared food ceramic plates, mugs and silverware to reduce the use of disposable items. There is also an active Buy Nothing Project network in Kingston, with three local chapters organized through Facebook groups and separated by geographic locations within the city. Members of these groups donate goods and services to each other but may only do so within the confines of their geography. The main objective is to reduce waste through reuse by finding new homes for articles that are no longer needed or wanted, or to borrow articles for short periods of time (tools for example). The City of Kingston also organizes two Giveaway Days, similar in philosophy to the Buy Nothing Project. On these designated days, households are allowed to place items by the curb for free pickup by other people. In 2019 Queen's University started a reusable food container initiative. Of the approximately 200,000 single-use containers that are normally used every semester, over 110,000 reusable containers were used in the

2022 fall semester, reducing the number of single-use containers by half (member of Queen's University Facilities, personal communication, February 22, 2023).

As these grassroots initiatives are proving popular, City officials are beginning to support them. Nonprofits can apply for small-grant funding through the annual Kingston Community Climate Action Fund, individuals can implement project ideas through the City's Neighborhood Climate Action Program, and information is disseminated through Sustainable Kingston's annual Climate Symposium as well as the Climate Action Kingston Newsletter. The "bottom up" approach is what Pollans et al. (2017) refer to as "visionary" because it prioritizes waste reduction through transformative action and emphasizes material reuse instead of the more energy-intensive mode of recycling. The limitations of these initiatives are that many of them are small in scale, and depend on volunteerism.

Conclusion

Despite public declarations committing to reducing waste, Kingston (like other cities across Canada) remains largely focused on the WH's bottom tiers—disposal and recycling (Figure 1.3). In its public statement on MSW reduction, the Government of Canada (2021) states that we "need to rethink how we purchase, use and throw out used items to reduce the costs and environmental impacts of waste management." But within this same statement, the federal government also states that we must view waste as a "valuable resource to increase the economic

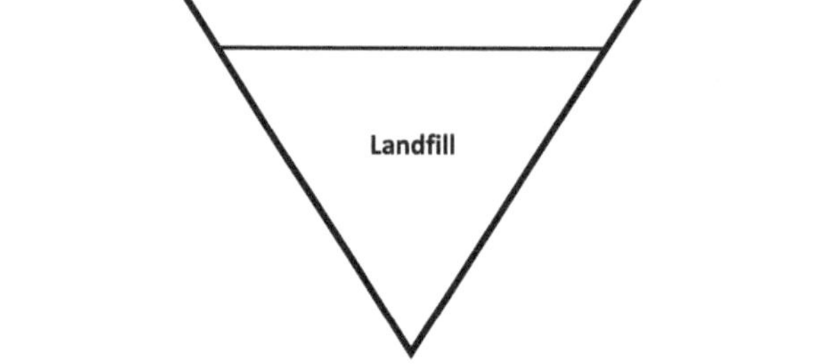

Figure 1.3 Kingston Waste Hierarchy diagram.

Note: Bars for Waste Prevention/Reduce, Remanufacture/Refurbish, and Energy Recovery are lacking because they do not factor into the governance of waste in Kingston. Area is correlated to waste management practices.

Source: Adapted from Zhang et al. (2022).

benefit from recycling and diversion" (Government of Canada, 2021). Several studies (e.g., Hird, 2022; Rocher, 2020) examine the contradictions between calls for reducing waste, and the continued emphasis on linear economics of both waste-to-energy and recycling. Waste-to-energy has been demonstrated to lead to increases in waste production (Hird, 2022), and is part of the linear economy, not the circular economy, as some claim, while recycling has both material limitations and environmental costs.

The City of Kingston continues to publicly emphasize waste reduction through processes that do not decrease waste generation. Not only do disposal and diversion fail to address the continued generation of waste, but both produce a positive feedback loop endemic to capitalist growth: increased waste and recycling generation leading to greater economic benefit to private companies. Since there are no legislative regulations or city targets for waste reduction, there are no citywide programs to collect, publish or disseminate this information. As one interviewee expressed, employees are only "legislatively required to provide and promote diversion programs" (KARC employee, personal communication, September 17, 2021).

While the federal government and municipalities such as Kingston remain largely compromised by a system that prioritizes private profit, individuals and grassroots organizations (along with a few small companies) are leading communities in waste reduction. Local, provincial and federal governments would do well to promote initiatives within the top tiers of the WH (reduce, reuse, and repair) through regulations and policies that emphasize the environment and human health, rather than economic benefit. With residents and small businesses already forging ahead through bottom-up initiatives to reduce waste generation, more could be accomplished to reduce waste if local, provincial and national levels of government would actualize their public ambitions to reduce waste through legislation.

Notes

1 Calculation was based on the production of 20,425 tons of disposal waste (including 13 percent of diverted waste which gets landfilled), using the average distance to the three landfills, one-way transportation of waste, 37 ton capacity of diesel trucks with 41.5 L/100 km fuel efficiency and transport emissions of 2263 g/L CO_2, 0.14 g/L CH_4, 0.082 g/L N_2O (Resource Productivity and Recovery Authority, 2024; Assamoi & Lawryshyn, 2012).

2 Brokers can be an important intermediary between the materials recovery facility (MRF) and recycling companies. Prices for commodities such as plastics, metals, glass, and paper are constantly changing, making it difficult for MRF managers to get the best return on their goods. Selling their packaged bales to brokers can improve profits, even though brokers claim a share of those profits.

References

Allen, D., Linsley, C., Spoelman, N., & Johl, A. (2024). *The fraud of plastic recycling*. Center for Climate Integrity.

Altman, R. (2021). On wishcycling. *Discard Studies*. Retrieved June 25, 2024 from https://discardstudies.com/2021/02/15/on-wishcycling.

Assamoi, B., & Lawryshyn, Y. (2012). The environmental comparison of landfilling vs. incineration of MSW accounting for waste diversion. *Waste Management*, 32: 1019–1030.

Blue Box Transition Schedule. (2024). *Resource productivity & recovery authority*. Retrieved September 10, 2024 from https://rpra.ca/wp-content/uploads/Blue-Box-Transition-Schedule-updated-Feb-2024.pdf.

Bulkeley, H., Watson, M., & Hudson, R. (2007). Modes of governing municipal waste. *Environment and Planning A*, 39(11): 2733–2753.

Canadian Council of Ministers of the Environment. (2009). *Canada-wide action plan for extended producer responsibility*. Retrieved June 26, 2004 from https://ccme.ca/en/res/cap-epr_e.pdf.

City of Kingston. (2024). *Waste strategies*. Retrieved June 14, 2024 from https://getinvolved.cityofkingston.ca/waste-strategies.

City of Kingston. (n.d.). *Waste strategies engagement summary*. Retrieved February 21, 2025 from https://ehq-production-canada.s3.ca-central-1.amazonaws.com/2e036a3e122e0fc70f2e6a02d798b910d440ed6b/original/1654103029/46998d26fb1a6538b46f0c1ebdc0ceab_Projects_WasteStrategiesEngagement_ReportingBack.pdf.

Dee, G. (2023). *Climbing up the waste hierarchy: The devil is in the details*. Masters thesis, Queen's University.

Environment and Climate Change Canada. (2022). *Canadian environmental sustainability indicators: Solid waste diversion and disposal*. Retrieved July, 7, 2022 from www.canada.ca/en/environment-climate-change/services/environmental-indicators/solid-waste-diversion-disposal.html.

Gille, Z. (2007). *From the cult of waste to the trash heap of history: The politics of waste in socialistic and postsocialist Hungary*. Indiana University Press.

Government of Canada. (2021). *Reducing municipal solid waste*. Retrieved August 17, 2022 from www.canada.ca/en/environment-climate-change/services/managing-reducing-waste/municipal-solid/reducing.html.

Government of Canada. (2023). *Budget 2023: A made-in-Canada-plan*. Retrieved May 2, 2023 from www.budget.canada.ca/2023/pdf/budget-2023-en.pdf.

Government of Canada. (2024a). *Municipal solid waste management in Canada*. Retrieved August 15, 2024 from www.canada.ca/en/environment-climate-change/services/managing-reducing-waste/municipal-solid/environment.html.

Government of Canada. (2024b). *Chapter 3: Lowering everyday costs*. Retrieved June 17, 2024 from www.budget.canada.ca/2024/report-rapport/chap3-en.html.

Government of Canada. (2024c) *Budget 2024: Fairness for every generation*. Retrieved August 15, 2024 from https://budget.canada.ca/2024/home-accueil-en.html#pdf.

Greenpeace. (n.d.). *Learn about plastic pollution*. Retrieved June 12, 2024 from www.greenpeace.org/international/campaigns/learn-about-plastic-pollution/#:~:text=The%20Recycling%20Myth,that%20converted%20back%20into%20packaging.

Hird, M. J. (2021). *Canada's waste flows*. McGill-Queen's University Press.

Hird, M. J. (2022). *A public sociology of waste*. Bristol University Press.

Hird, M. J., & Kuyvenhoven, C. (2020). Waste transportation and the downstream-recycle, upstream-reduce equation: Waste's management within neoliberal capitalism. In R. Ek and N. Johannson (Eds.), *Perspectives on waste from the social sciences and humanities: Opening the bin*, Cambridge Scholars Press, pp. 79–97.

Hird, M. J., Lougheed, S., Rowe, R. K., & Kuyvenhoven, C. (2014). Making waste management public (or falling back to sleep). *Social Studies of Science*, 44: 441–465. https://doi.org/10.1177/0306312713518835.

Kawai, K., & Tasaki, T. (2016). Revisiting estimates of municipal solid waste generation per capita and their reliability. *Journal of Material Cycles and Waste Management*, 18: 1–13. https://doi.org/10.1007/s10163-015-0355-1.

Kaza, S., Yao, L., Bhada-Tata, P., & Van Woerden, F. (2018). *What a waste 2.0: A global snapshot of solid waste management to 2050.* Urban Development Series. World Bank. https://doi.org/10.1596/978-1-4648-1329-0.

Lansink, A. (2017). *Challenging changes: Connecting waste hierarchy and circular economy.* LEA Nijmegen.

Lougheed, S., Hird, M. J., & Rowe, R. K. (2016). Governing household waste management: An empirical analysis and critique. *Environmental Values*, 25(3): 287–308.

Macbride, S. (2013). *Recycling reconsidered. The present failure and future promise of environmental action in the United States.* The MIT Press.

Ontario Waste Management Association. (2021, January). *State of waste in Ontario: Landfill report.* Retrieved February 21, 2025 from https://pub-wellington.escribemeetings.com/filestream.ashx?DocumentId=593

Pollans, L. B. (2017). Trapped in trash: Modes of governing and barriers to transitioning to sustainable waste management. *Environment and Planning*, 49(10): 2300–2323. https://doi.org/10.1177/0308518X17719461.

Province of Ontario. (2021). *Renewing the blue box: Final report on the blue box mediation process.* Retrieved June 25, 2024 from www.ontario.ca/page/renewing-blue-box-final-report-blue-box-mediation-process

Resource Productivity and Recovery Authority. (2024). *Datacall.* Retrieved June 5, 2024 from https://rpra.ca/programs/about-the-datacall/#:~:text=of%20Municipal%20Groups-,DATACALL%20REPORTS,-Consolidated%20Datacall%20Report.

Rocher, L. (2020). Waste, a matter of energy, a diachronic analysis (1992–2017) of waste-to-energy rationales. In *Perspectives on waste from the social sciences and the humanities. Opening the Bin.*

Sawell, S. E., Hetherington, S. A., & Chandler, A. J. (1996). An overview of municipal solid waste management in Canada. *Waste Management*, 16(5/6): 351–359.

Statistics Canada. (2021a). *Table 38-10-0033-01. Materials diverted, by source, inactive* (data table). https://doi.org/10.25318/3810003301-eng.

Statistics Canada. (2021b). *Table 38-10-0036-01 Local government characteristics of the waste management industry* (data table). https://doi.org/10.25318/3810003601-eng.

Statistics Canada. (2022a). *Table 17-10-0005-01. Population estimates on July 1st, by age and sex* (data table). https://doi.org/10.25318/1710000501-eng.

Statistics Canada. (2022b). *Table 38-10-0138-01. Waste materials diverted, by type and by source* (data table). https://doi.org/10.25318/3810013801-eng.

Statistics Canada. (2022c). *Focus on geography series, 2021 Census of population, Kingston, Census Metropolitan Area.* Retrieved 16 August, 2024 from www12.statcan.gc.ca/census-recensement/2021/as-sa/fogs-spg/page.cfm?topic=1&lang=E&dguid=2021S0503521.

Statistics Canada. (2023). *Table 38-10-0032-01. Disposal of waste, by source* (data table). https://doi.org/10.25318/3810003201-eng.

United Nations. (2022). *The Sustainable Development Goals report.* Retrieved 30 September 2023 from https://unstats.un.org/sdgs/report/2022.

United Nations Environment Programme. (2012). *The global garbage crisis: No time to waste.* Retrieved 12 October 2022 from www.unep.org/news-and-stories/press-release/global-garbage-crisis-no-time-waste.

United Nations Environment Programme. (2024). *Global waste management outlook 2024*. Retrieved 15 August 2024 from www.unep.org/resources/global-waste-management-outlook-2024.

Wilkinson, S. (2020). *When will Canadians have the right to repair?* Canadian Centre for Policy Alternatives. Retrieved 2 May 2023 from https://policyalternatives.ca/publications/monitor/when-will-canadians-have-right-repair#:~:text=In%20part%20inspired%20by%20activism,replacement%20parts%2C%20software%20and%20other.

Zhang, C., Hu, M., Di Maio, F., Sprecher, B., Yang, X., &Tukker, A. (2022). An overview of the waste hierarchy framework for analyzing the circularity in construction and demolition waste management in Europe. *Science of the Total Environment*, 803. https://doi.org/10.1016/j.scitotenv.2021.149892.

Zink, T., & Geyer, R. (2019). Material recycling and the myth of landfill diversion. *Journal of Industrial Ecology*, 25(3): 541–548. https://doi.org/10.1111/jiec.12808.

2 Municipal strategies for zero waste planning

Thoreau Rory Tooke and Christina Seidel

Introduction

Cities are increasingly grappling with environmental challenges and taking measures to address climate change, promote sustainability, and protect the natural environment (Riffat et al., 2016). Among the broad range of environmental initiatives, solid waste has some of the longest and most direct ties to community issues and local governmental responsibility. Traditionally, municipalities across North America have acted as service providers for the collection and disposal of waste generated within their communities, and as stewards of the local environment; cleaning litter and debris from roadways, sidewalks, parks and beaches.

Modern environmental sustainability requires that municipalities expand their role as solid waste service providers to encompass waste reduction, reuse and circularity. These concepts themselves are not new, but the language and terminology has evolved over time, with the philosophy of *zero waste* presenting a framing and set of practices that are both broadly encompassing of environmental sustainability while remaining focused on solid waste management and its established system of actors, institutions and services (Valenzuela-Fernández & Escobar-Farfán, 2022).

Zero waste is a goal to guide continuous improvements to the waste management system and is often linked with the paradigm of a circular economy that couples economic wellbeing with environmental sustainability. Global society has generally succeeded in a relative decoupling of economic growth from material use, meaning that resource use has increased less than Gross Domestic Product (Haberl et al., 2020). However, the environmental harm associated with the current global economy cannot be mitigated without an absolute reduction in resource use (Haberl et al., 2020). The intentional coupling of the economy and environment through a circular economy promises to be one of the most significant social shifts in the sustainable management of the environment to emerge in decades.

A comprehensive assessment of the economic impacts of a global shift to a circular economy has not been determined, but the focus on plastics as an especially ubiquitous and problematic material for the environment hints at some of these effects. Modeling in support of the United Nations Intergovernmental Negotiating Committee on Plastic Pollution indicates that economic activity is comparable between business-as-usual and a policy scenario that would establish

DOI: 10.4324/9781003534129-4

absolute limits in plastic production and consumption and eliminate problematic plastics from the economy (Systemiq, 2024). Under this global scenario, up to $600 billion of economic activity would be shifted from primary plastic production, waste collection and disposal to new value pools in recycling, plastic alternatives and reuse; leading to a more diversified and sustainable economy (Systemiq, 2024).

Shifting to zero waste is also being recognized as an effective strategy to mitigate climate change. The World Trade Organization (2021) has assessed that the production and transportation of internationally traded goods and services accounts for 20–30% of global greenhouse gases. Municipalities are starting to account for these emissions in their greenhouse gas inventories as "embodied emissions" that encompass the emissions generated from resource extraction, production and transportation of the products and materials used and consumed within their local economies. In the case of the City of Victoria (which is referenced later in this chapter as a case for describing zero waste planning strategies) the embodied emissions have been assessed to be 2.5 times greater than the territorial emissions generated from the direct combustion of fossil fuels for transportation and heating (City of Victoria, 2024a). Thirty percent of these embodied emissions are attributed to products and waste (City of Victoria, 2024a).

Building from the strong foundation that local governments have as waste collection and disposal service providers and expanding their role to support zero waste, municipalities are well positioned to advance the paradigm of the circular economy in support of a more resilient local economy that will achieve a broad range of local and global environment benefits. This chapter describes robust planning strategies that are grounded in evidence to support this endeavor. The intention of this chapter is also to offer a blueprint for other communities embarking on similar processes and to use Zero Waste Victoria as an example of a strategic and comprehensive approach to planning and implementation. Finally, through this chapter, the authors hope to continue and grow a dialogue between academia and practitioners that will advance the long-term sustainability of our communities.

Beyond recycling

It is well understood that recycling is not a complete solution to address the environmental impacts of waste management. The concept of zero waste arose from desires to reduce the impacts associated with waste, and was led by groups such as the Zero Waste International Alliance that was formed in 2003 to "promote positive alternatives to landfill and incineration and to raise community awareness of the social and economic benefits to be gained when wasted materials are regarded as resources which can create both employment and business opportunities" (Zero Waste International Alliance, 2021).

Since that time, waste policy has continued to evolve, with many communities in Canada now embracing circular economy strategies (Circle Economy, 2024). Circular economy is a more holistic approach that advances a

> system where materials never become waste and nature is regenerated. In a circular economy, products and materials are kept in circulation through processes like maintenance, reuse, refurbishment, remanufacture, recycling, and composting. The circular economy tackles climate change and other global challenges, like biodiversity loss, waste, and pollution, by decoupling economic activity from the consumption of finite resources.
>
> (Ellen MacArthur Foundation, 2024)

In pursuit of this system, the circular economy is based on three principles, driven by design (Ellen MacArthur Foundation, 2024):

- Eliminate waste and pollution
- Circulate products and materials (at their highest value)
- Regenerate nature

In terms of the difference between zero waste and circular economy, while zero waste focuses on reducing waste, circular economy is broader and goes one step further by striving to regenerate the environment. To summarize, zero waste is a set of principles, while a circular economy is a model that provides a systematic framework. Essentially, zero waste is a goal, that may be a component of circular economy, whereas circular economy is a means to get there (Zero Waste International Alliance, 2021).

In assessing the environmental impact of waste management options, the waste management hierarchy is often cited. The hierarchy presents options in a framework that provides preference to practices higher on the hierarchy, following the 4R's order: Reduction, Reuse, Recycling and Recovery, followed by Disposal, as shown in Figure 2.1 (Recycling Council of Alberta, 2014).

This has followed through to ISO-14001; an internationally recognized standard for environmental management systems that provides guidelines for organizations to establish, implement, and maintain effective environmental management practices (International Organization for Standardization, 2015). The standard requires organizations to prioritize sustainable waste management practices, including using the waste hierarchy, as follows (International Organization for Standardization, 2015):

Step 1: Prevention

Eliminate the need for waste management processes by avoiding the production of materials, using less material in the design and manufacture of products, and designing products for longer use.

Step 2: Preparation for reuse

Cleaning, repairing or refurbishing items to extend their lifespan and prevent them from becoming waste.

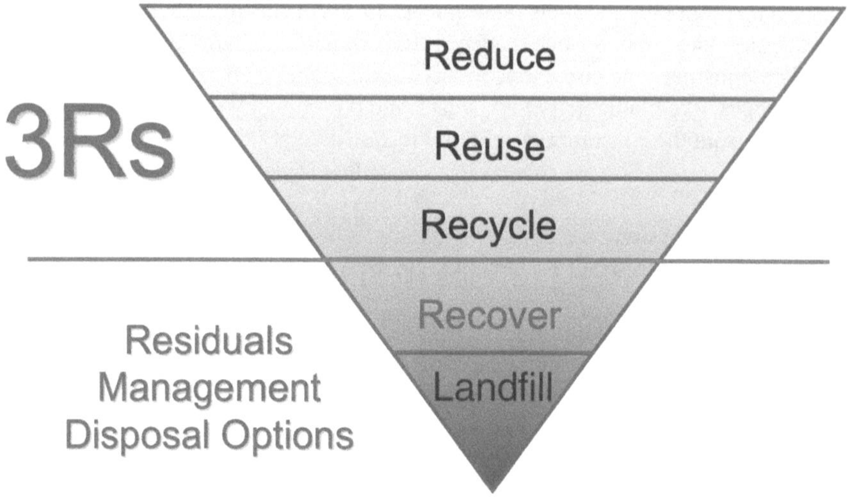

Figure 2.1 Waste management hierarchy.

Source: This waste management hierarchy is adopted from the Recycling Council of Alberta.

Step 3: Recycling

Converting waste materials into new products (including composting).

Step 4: Recovery

Accessing the embodied energy from waste that cannot be reused or recycled.

Step 5: Disposal

Sending waste to landfill or incineration when no other options are available.

To quantify the impact of waste management options, the United States Environmental Protection Agency (USEPA, 2024) created the Waste Reduction Model to provide high-level comparisons of potential greenhouse gas emissions reductions, energy savings, and economic impacts when considering different materials management practices, including source reduction, recycling, composting, combustion and landfilling.

The relative greenhouse gas emissions of various diversion options in Canada are shown in Figure 2.2 (ICF Consulting, 2005). This was performed by ICF Consulting using similar methodology to that used by United States Environmental Protection Agency for the WARM model (USEPA, 2024), while incorporating Canadian data.

The negative numbers in Figure 2.2 represent carbon benefits, where greenhouse gas impacts are lower than the business-as-usual scenario. As can be seen,

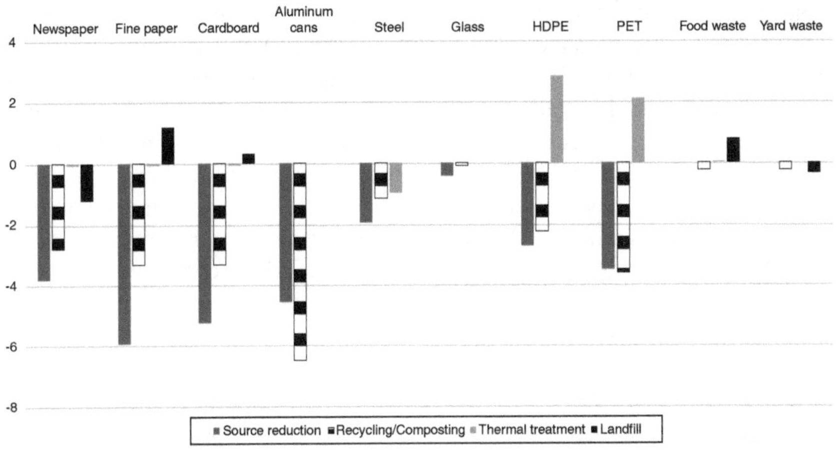

Figure 2.2 Greenhouse gas impacts of waste management activities in Canada (tons eCO_2/tons material managed).

the relative benefits almost invariably follow the waste management hierarchy, substantiating and supporting policies to prioritize options higher on the hierarchy. For example, despite the clear greenhouse gas benefits of recycling, source reduction offers greater reductions in all cases except aluminum and PET (Figure 2.2). It is also important to point out that the recycling benefits assume full recycling capture rates, that are difficult to achieve in practice.

These conclusions also validate the initial assertion of this section that recycling does not offer a full answer to minimizing the environmental impacts of waste management. Instead, it is important to work to capture the strong benefits of source reduction and reuse when developing waste policy. This is despite the tendency to look to enhanced municipal material collection services and policy approaches like extended producer responsibility that rely heavily on recycling as a key outcome. Instead, the generally more significant benefits associated with source reduction (including reuse) need to be captured through more holistic policy initiatives that may be offered through approaches like circular economy that embrace approaches like source reduction, reuse and refurbishment (Ellen MacArthur Foundation, 2024). Only by addressing these higher-level approaches, including reducing the primary generation of waste through actions that decouple waste generation from economic activity, can we start to fundamentally reduce the environmental impacts associated with waste management practices.

Strategic planning and reporting cycles

A planning imperative arises as municipalities move from discrete provision of solid waste services toward the goal-oriented community-wide changes inherent in zero waste. Zero waste is a long-term endeavor and planning needs to consider the timelines and sequence of activities that will support sustained waste reduction.

On a longer-term horizon, a plan should be a strategic blueprint that remains relevant throughout the designated planning timeframe, remaining "evergreen" in its approach by providing a clear and achievable framework that can be sustained even as external conditions change. The long-range plan should be a guide for actions, but care should be taken not to establish or commit to discrete actions. A fundamental key to successful long-range planning is the ability to adapt to change (Copeland et al., 2021). At the same time, the plan needs to align with the community's realities and readiness for change, reflecting and communicating a clear understanding of the values, identity and environment held by the stakeholders who will be involved in the changes (Leal Filho et al., 2022).

While a long-range plan sets the framework and guides the direction to zero waste, it is also important to have shorter-term plans that specify the actions and resources needed to achieve the longer-term goals. A short-term plan provides an opportunity to determine discrete actions in response to the current conditions of the community. Short-term planning should take advantage of opportunities presented in the market or through new technologies and adjust previous initiatives based on new information. Coupling progress reporting and short-term action planning provides an opportunity to transparently describe the impacts of past actions and present a stronger rationale for future initiatives.

A strategic approach to the timing of short-term planning is important for success. The concept of a "policy window" (Kingdon, 2010) provides a useful framing for the strategic application of short-term zero waste planning by considering opportunities for intervention when: a problem arises; a policy solution is available; and there is political will for action. Developing an adaptable long-range plan can create the context for establishing set intervals of short-term planning but requires that current issues and potential solutions continue to be examined and understood across planning cycles. In the municipal context, the short-term plan is ideally timed with Council election cycles to identify alignment with Council priorities or strategic planning initiatives that would support attention to and accountability for the short-term plan.

Designing a planning process at the outset with the intent to integrate the long-range plan with short-term action planning cycles (Figure 2.3) promotes the ongoing use and relevance of the plan while providing a strong mandate for implementation.

Stakeholder engagement and relationships

As with all public decision-making processes, identifying and engaging stakeholders is an integral and critical part of zero waste planning. It is important that stakeholder engagement be proactive by being integrated early in the planning process to gauge community feedback to proposed strategies and identify potential barriers to success. Early engagement can also serve to identify community champions, or ambassadors, who can provide leadership and mentorship that will assist in advancing waste reduction approaches. Community champions programs are often the key to successfully empowering and supporting emergent leaders in a community

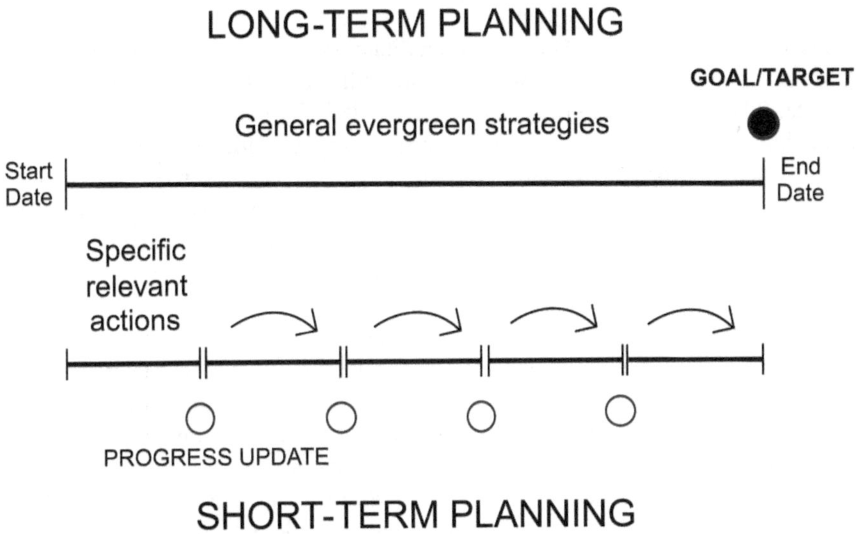

Figure 2.3 Long-term and short-term waste management planning.

Note: Conceptual representation of the approach and integration of long-range objectives and strategies with short-term action planning cycles.

to help promote programs like waste reduction (Center for Scientific Collaboration and Community Engagement, 2024).

When applied specifically to waste reduction, where behavior change is integral to success, understanding the motivations and behaviors of different stakeholders is essential to targeting the right messages to the right audiences (Waste and Resources Action Programme, 2014). Stakeholder engagement can take many approaches, with successful methods including online surveys, community pop-up events, town hall sessions, open engagement sessions, targeted focus groups and interviews (Intelligent Futures, 2024; Strategy Management Consulting, 2024). The choice of format needs to consider issues such as accessibility, where online approaches can offer advantages, contrasted with a desire for greater insight into attitudes regarding specific concepts, where in-person sessions such as focus groups are preferred (Strategy Management Consulting, 2024). It is important that the engagement process is seen as transparent and authentic, providing stakeholders with a perceived genuine opportunity to shape the final plan. A well-designed process also assesses the community's readiness for required changes, pinpointing specific, actionable steps that need to be included in subsequent action plans.

Engagement is used to solicit input on strategies, facilitating the addition, removal, or modification of strategic elements as dictated by stakeholder feedback. Stakeholder consultation can help identify the opinions and expectations of various sectors key to program outcomes through providing a forum for feedback on potential programming options. Stakeholder feedback is then incorporated into subsequent strategy development.

Public consultation can also serve to facilitate program implementation through the role it can play in public awareness and education (Rural Municipalities of Alberta, 2018). Since the engagement process is important to program development, how to choose stakeholders becomes a key consideration. Generally, in public processes, the general public is invited to participate in stakeholder engagement through some inclusive approach like online surveys. However, there are good reasons to target specific stakeholders, based on their legitimacy to the process. Legitimacy can be examined using the following elements (Bammer, 2021):

1 Potential value of contributions: Stakeholders who bring extensive experience or insights into the issue at hand have greater legitimacy than those with shallow or peripheral viewpoints. This is also where leaders and community champions can be identified and prioritized for the potential value they can provide.
2 Stakeholder's interests: Stakeholders who work for the good of a group or society have greater legitimacy than those who are purely self-interested.
3 Representation: Formal representatives of stakeholder groups have more legitimacy than those acting in self-interest.

It is important to recognize that the invitation to participate in the planning process can confer legitimacy on stakeholders and their perspectives. This can be used deliberately to increase the legitimacy of marginalized groups, but care must be taken to recognize the potential influence on resulting outcomes (Bammer, 2021).

Legitimacy may not be uniform across various elements and there is no easy way to weigh the importance of different aspects when selecting stakeholders. Therefore, efforts should be made for stakeholder selection to be undertaken in an unbiased, neutral way using approaches like choices by groups of selectors rather than individuals.

Target setting and progress measurement

Target setting is a common practice for solid waste planning and policy development. Setting a target often acts as the catalyst for change and provides elected officials the opportunity to demonstrate commitment to environmental outcomes.

The Government of British Columbia provides solid waste planning guidelines intended for use by regional districts who are obligated to develop plans for the management of municipal solid waste and recyclable materials (British Columbia Ministry of Environment, 2016). Under these guidelines, target setting is recommended as one of the initial steps which is then followed by the development of strategies and public consultation before finalizing a plan. In practice, this sequence of steps can lead to a situation of high ambition but requires a confident understanding of the impacts of the chosen strategies and ongoing political support to ensure robust implementation.

Waste reduction targets ought to be considered strategically, playing a role not only at the outset of plan development, but at various points in the planning process and throughout implementation. Waste reduction targets are commonly set early in planning processes and intended to guide future decision-making. In practice, planning processes often do not account for the rigor, time and resources to quantify the impacts of specific decisions for achieving a specific target, which often leads to a final plan that falls short of its initial ambition. Although galvanizing support for a target early can ideally lead to the prioritization of options and actions based on their impact, another of the major pitfalls of this approach is the potential alienation of stakeholders and political support over time. Specifically, community champions whose contributions could be considered low impact may become distanced from implementation and compromise political will. Setting an ambitious target that cannot realistically be achieved with a given set of options may also compromise success as decisionmakers start to perceive that actions do not demonstrably lead to progress.

The waste reduction target ought to be achievable and determined based on a sound and transparent approach that identifies the impacts of the interventions being proposed in the plan. As a result, there is an advantage to establishing a target in the latter stages of the planning process, once options have been identified and stakeholder support has been confirmed (see Figure 2.4). During the implementation phase of the plan the target should also act as a progress indicator and communicated at strategic intervals, such as the short-term planning cycles noted above. Reporting progress is not only an important mechanism for transparency, but behavioral science studies have consistently highlighted that achieving interim targets can maintain commitment and momentum toward longer-term objectives (Locke & Latham, 2002; Gollwitzer, 1999; Fishbacj & Dhar, 2005; Huang & Zhang, 2013). This also reinforces the strategic value of establishing a realistic, achievable and technically determined target as opposed to an aspirational target.

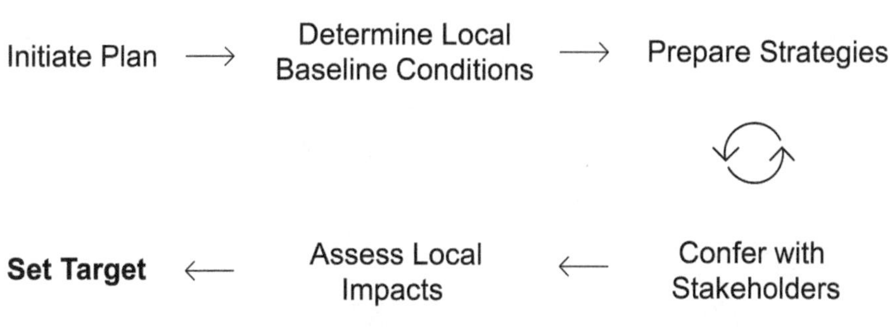

Figure 2.4 Target setting.

Note: This represents the recommended approach to target setting where strategies are prepared with stakeholders and the target established at the end of the planning process.

Zero Waste Victoria

The City of Victoria is the capital of the province of British Columbia and the economic and institutional hub of Southern Vancouver Island. Victoria is home to roughly 100,000 people in 2024 with two-thirds of its population living in multifamily buildings, while the daytime population often doubles with workers and visitors concentrating in the downtown core of the city. The City of Victoria provides waste collection services to ground-oriented homes and is responsible for cleaning litter from streets and parks, while the private sector manages the remainder of the material from multifamily, commercial, institutional and industrial properties as well as from construction, renovation and demolition activities. In 2020, approximately 90% of the material that moved through the city was outside the direct control of municipal solid waste services (City of Victoria, 2020).

The plastic pollution crisis acted as a catalyst and the emerging paradigm of the circular economy acted as a framework for the City to investigate its role in supporting a more responsible approach to managing all materials across the community. After two years in development, Zero Waste Victoria was adopted in 2020 as the City's first long-term plan for waste reduction; setting a 50% decrease in landfilled waste by 2040 as its target (City of Victoria, 2020).

The planning process for Zero Waste Victoria began with an assessment of the sources and composition of material generated within city boundaries (City of Victora, 2020). Landfill waste composition studies provided the primary source of information for the City's material inventory but required the disaggregation of regionwide data by sector to capture Victoria's share of waste. Materials diverted to recycling and composting were also integrated along with landfilled waste to provide a full understanding of the flow of materials through the city. The outcome of the assessment is shown in Figure 2.5 and demonstrates that the residential,

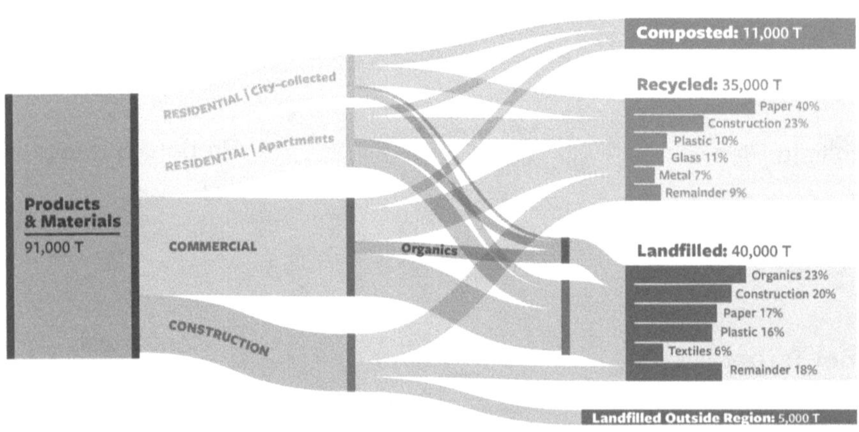

Figure 2.5 Representation of material flows by sector and material for the City of Victoria in 2020. This figure is reproduced with permission from the City of Victoria.

commercial and construction sectors are the primary generators of waste and that half of the materials are currently diverted for recycling or composting. The figure also shows that the highest diversion (72%) is achieved in the residential sector served by the City's curbside collection services and the lower diversion from apartments and the commercial sector is largely due to the higher portion of organics being landfilled from these sectors.

These insights from the assessment shown in Figure 2.5 provided a springboard for discussion and proved a critical step for the plan's success by creating a sector-specific baseline to benchmark performance; identifying target stakeholders and materials; and recognizing the function and capacity of the established material diversion system.

An initial set of strategies addressing the most prevalent and challenging materials was informed by the comprehensive analysis of material flows within the city. These 40 strategies were developed to align with the established waste prevention hierarchy with reduction and reuse initiatives given priority. Strategies were considered feasible if they were within the general authority of local governments and included service design and delivery, regulatory measures, and procurement (primarily related to capital projects such as infrastructure and public facilities). Education, while not identified as a standalone strategy due to the evidence that in isolation it does not lead to significant change (McGuire, 2015; Heimlich & Ardoin, 2008), was recognized from a behavioral perspective as a supportive element of any successful policy, program, or service.

The drafted strategies were then used as part of a six-month engagement process. The engagement was designed to gather six separate groups of stakeholders from similar sectors (environmental groups, businesses and industry associations, construction, government, solid waste service providers, and neighborhood associations) in a focus group format in addition to several broad open engagement sessions. The engagement was used to solicit feedback on the strategies, facilitating the addition, removal, or modification of strategies as necessary. This approach ensured stakeholders were provided an authentic opportunity to shape the final plan while also assessing the community's readiness for change and pinpointing specific, actionable steps to include in the first short-term action plan.

Once the strategies were confirmed with the stakeholders the waste reduction target was established. The target was developed through an exercise that examined each strategy in terms of best practice interventions, existing precedents, known limitations and system constraints and then extrapolating the potential waste reduction or diversion impact for each material category as applied to the baseline assessment for Victoria. To do this, every strategy was linked to any relevant material category from the landfill composition study (total of 90 categories) that may be impacted by the intervention and then identified a maximum reduction or diversion level using known precedents and theoretical limits. For example, the City of Victoria's experience regulating plastic checkout bags indicated that 90% of these materials from the residential sector could be eliminated or diverted. In contrast, no commercially viable diversion options were identified for vinyl flooring and therefore the analysis assumed all this material would continue to be landfilled.

Following from this technical exercise, an achievable target was determined to be a 50% decrease in waste sent to landfill over a 20-year period.

Since Zero Waste Victoria was adopted in 2020 it has gone through two short-term action planning cycles. The first short-term plan was adopted at the same time as the long-range plan and included 15 actions that were completed over a three-year period. The second plan was adopted in 2024 with 13 new actions to be implemented over a four-year period. By 2027, it is expected that the City of Victoria's actions will have reduced annual waste disposal by 9,200 metric tons, progressing the City 41% of the way to its 2040 target (City of Victoria, 2024b).

Conclusion

Zero waste is a goal that guides ongoing waste management system improvements and in practice is closely tied within the circular economy concept. Circular economy, which deliberately links economic wellbeing to environmental health, is likely to be one of the biggest movements in decades for the sustainable management of waste in our society. With the opportunity for a host of advantages for the local and global environment, municipalities ought to consider their role in planning for a shift to zero waste and advancing the circular economy in their communities.

Regardless of intentions, without a plan and an approach to implementation, municipalities will be left in a position where they are reacting to environmental degradation with only traditional waste collection and disposal services that will grow increasingly outdated as an effective or efficient solution to modern community needs. A planning process that integrates long-range objectives with short-term action planning is recommended as the best opportunity for ensuring ongoing implementation that can be tied to Council election cycles to regularly open political "windows" for policy interventions.

Carefully selected community stakeholders with an authentic opportunity to inform the plan are key to the success of the process, including finding alignment for the long-term objectives and informing the strategies to meet the objectives. Stakeholder engagement can also capture and commit stakeholders to initial actions that can begin at the same time as the plan adoption to ensure momentum is generated early.

A technically derived target should be established at the end of the planning process, once the plan and strategies have been developed and confirmed with the community. This offers an approach grounded in behavior science to maintain implementation momentum while also being realistic, transparent, and defensible.

Zero Waste Victoria provides an example of a strategic and comprehensive approach to planning and implementation, setting a 50% decrease in landfilled waste by 2040 as its target, with an initial set of strategies addressing the most prevalent and challenging materials developed to align with the established waste management hierarchy.

Engagement needs to ensure stakeholders are provided an authentic opportunity to shape the final circular economy plan and deliver a waste reduction target that encompasses all the streams of material generated in the community.

Zero Waste Victoria has maintained its implementation momentum and demonstrated impact and success, realizing some initial benefits of a circular economy for its community, while on track to achieving its 2040 target.

References

Bammer, G. (2021). *Stakeholder engagement primer 3: selecting stakeholders.* I2Insights. https://i2insights.org/2021/10/28/selecting-stakeholders.

British Columbia Ministry of Environment. (2016). *A guide to solid waste management planning.* www2.gov.bc.ca/assets/gov/environment/waste-management/garbage/swmp.pdf.

Center for Scientific Collaboration and Community Engagement. (2024). *Community champions programs.* www.cscce.org/resources/community-champions-programs.

Circle Economy. (2024). *Canada's Circular Cities and Regions Initiative is leading the circular economy transition.* www.circle-economy.com/news/canadas-circular-cities-and-regions-initiative-is-leading-the-circular-transition#:~:text=Ten%20more%20communities%20have%20now,Granby%2C%20Outremont%2C%20and%20Chibougamau.

City of Victoria. (2020). *Zero Waste Victoria.* https://pub-victoria.escribemeetings.com/filestream.ashx?DocumentId=62407.

City of Victoria. (2024a). *Climate Leadership Program update.* https://pub-victoria.escribemeetings.com/filestream.ashx?DocumentId=96527.

City of Victoria. (2024b). *Zero waste update and four year action plan.* https://pub-victoria.escribemeetings.com/filestream.ashx?DocumentId=95908.

Copeland, L., Littlecott, H., Couturiaux, D., Hoddinott, P., Segrott, J., Murphy, S., Moore, G., & Evans, R. (2021). The what, why and when of adapting interventions for new contexts: A qualitative study of researchers, funders, journal editors and practitioners' understandings. *PLoS ONE*, 16(7). https://doi.org/10.1371/journal.pone.0254020.

Ellen MacArthur Foundation. (2024). *What is a circular economy?* www.ellenmacarthurfoundation.org/topics/circular-economy-introduction/overview.

Fishbach, A., & Dhar, R. (2005). Goals as excuses or guides: The liberating effect of perceived goal progress on choice. *Journal of Consumer Research*, 32(3), 370–377. https://doi.org/10.1086/497548.

Gollwitzer, P. M. (1999). Implementation intentions: Strong effects of simple plans. *American Psychologist*, 54(7), 493–503. https://doi.org/10.1037/0003-066X.54.7.493.

Haberl, H., Wiedenhofer, D., Virág, D., Kalt, G., Plank, B., Brockway, P., Fishman, T., Hausknost, D., Krausmann, F., Leon-Gruchalski, B., Mayer, A., Pichler, M., Schaffartzik, A., Sousa, T., Streeck, J., & Creutzig, F. (2020). A systematic review of the evidence on decoupling of GDP, resource use and GHG emissions, part II: Synthesizing the insights. *Environmental Research Letters*, 15(6). https://doi.org/10.1088/1748-9326/ab842a.

Heimlich, J. E., & Ardoin, N. M. (2008). Understanding behavior to understand behavior change: A literature review. *Environmental Education Research*, 14(3), 215–237. https://doi.org/10.1080/13504620802148881.

Huang, S. C., & Zhang, Y. (2013). Motivational consequences of perceived velocity in consumer goal pursuit. *Journal of Marketing Research*, 50(6), 1063–1075.

ICF Consulting. (2005). *Determination of the impact of waste management activities on greenhouse gas emissions, 2005 update*. https://doi.org/10.2307/23142903.
Intelligent Futures. (2024). www.intelligentfutures.ca.
International Organization for Standardization. (2015). *ISO 14001:2014 environmental management systems: Requirements with guidance for use* (3rd ed.). www.iso.org/standard/60857.html.
Kingdon, J. (2010). *Agendas, alternatives and public policy* (2nd ed.). Pearson.
Leal Filho, W., Levesque, V., Sivapalan, S., Salvia, A. L., Fritzen, B., Deckert, R., Kozlova, V., LeVasseur, T. J., Emblem-Perry, K., Azeitero, U. M., Paco, A., Borsari, B., & Shiel, C. (2022). Social values and sustainable development: Community experiences. *Environment Sciences Europe*, 34(1), 67. https://doi.org/10.1186/s12302-022-00641-z.
Locke, E. A., & Latham, G. P. (2002). Building a practically useful theory of goal setting and task motivation: A 35-year odyssey. *American Psychologist*, 57(9), 705–717. https://doi.org/10.1037/0003-066X.57.9.705.
McGuire, N. M. (2015). Environmental education and behavioral change: An identity-based environmental education model. *International Journal of Environmental & Science Education*, 10(5), 695–715. https://doi.org/10.12973/ijese.2015.261a.
Recycling Council of Alberta. (2014). *Recycling 101*. www.recycle.ab.ca.
Riffat, S., Powell, R., & Aydin, D. (2016). Future cities and environmental sustainability. *Future Cities and Environment*, 2(1). https://doi.org/10.1186/s40984-016-0014-2.
Rural Municipalities of Alberta (2018). *Public engagement guide*. https://rmalberta.com/wp-content/uploads/2018/10/RMA-AUMA-Public-Engagement-Guide.pdf.
Strategy Management Consulting. (2024). *What is stakeholder engagement, and why is it important for strategic planning?* www.smestrategy.net/blog/stakeholder-engagement-management-for-strategic-planning.
Systemiq. (2024). *Plastic treaty futures*. www.systemiq.earth/reports/downloads/Systemiq-Plastic_Treaty_Futures_EN.pdf.
United States Environmental Protection Agency. (2024). *Basic information about the waste reduction model*. www.epa.gov/warm/basic-information-about-waste-reduction-model.
Valenzuela-Fernández, L., & Escobar-Farfán, M. (2022). Zero-waste management and sustainable consumption: A comprehensive bibliometric mapping analysis. *Sustainability*, 14(23), 16269. https://doi.org/10.3390/su142316269.
Waste and Resources Action Programme. (2014). *Household waste prevention hub: Delivering waste prevention—Stakeholders*. www.wrap.ngo/resources/guide.
World Trade Organization. (2021). *The carbon content of international trade*. www.wto.org/english/news_e/news21_e/clim_03nov21-4_e.pdf.
Zero Waste International Alliance. (2021). *Zero waste vs circular economy: Your guide to getting in the loop*. https://zwia.org.

3 Challenges and opportunities for industrial, commercial and institutional (IC&I) waste compliance in Nova Scotia

Hunar Arora and Tony R. Walker

Introduction: Waste management in Nova Scotia

Nova Scotia (NS) is committed to maintaining a robust economy and a healthy environment. It has acknowledged that solid waste represents a valuable resource, capable of generating employment opportunities through efficient and environmentally conscious management practices (Government of Nova Scotia, 2023). It believes that regional collaboration can minimize the escalation of capital and operational expenses linked to meeting environmental criteria for waste disposal and attaining greater levels of waste diversion. NS is divided into seven waste management regions:

- Region 1 is known as the Cape Breton Region, comprised of the counties of Cape Breton, Inverness, Victoria and Richmond;
- Region 2 is known as the Eastern Region, comprised of the counties of Antigonish, Pictou and Guysborough;
- Region 3 is known as the Northern Region, comprised of the counties of Colchester and Cumberland and the District of East Hants;
- Region 4 is known as the Halifax Region, comprised of the county of Halifax;
- Region 5 is known as the Valley Region, comprised of the counties of Annapolis and Kings;
- Region 6 is known as the South Shore Region, comprised of the counties of Lunenburg and Queens and the District of West Hants;
- Region 7 is known as the Western Region, comprised of the counties of Digby, Shelburne and Yarmouth.

Waste in NS is categorized into two categories: residential waste; and nonresidential waste, such as industrial, commercial and institutional (IC&I) waste. Residential waste includes waste from primary and seasonal dwellings, which includes all single-family, multi-family, high-rise and low-rise residences (Statistics Canada, 2024). This waste is usually collected by the municipalities or self-hauled to depots, transfer stations or landfills. Nonresidential or IC&I waste in NS is further

DOI: 10.4324/9781003534129-5

categorized into industrial waste being generated by manufacturing, and primary and secondary industries including the Construction & Demolition (C&D) industry; commercial waste generated by commercial operations and institutional waste generated by institutional facilities (Statistics Canada, 2024). IC&I waste can sometimes be picked up by municipalities or through contracts with private hauling companies, depending upon the regulations of each region (S. Betts, personal communication, July 11, 2023).

Waste management regulations and policies in NS includes Solid Waste Resource Management Strategy, Guidelines for Municipal Waste Transfer Facility, Materials Banned from Disposal Sites and Extended Producer Responsibility (EPR) for Packaging, Paper Products and Packaging-Like Products Regulations (Diggle & Walker, 2020; Diggle et al., 2023). Recently, the provincial government announced its target to reduce solid waste generation to 300 kg/person/year by 2030 (Government of Nova Scotia, 2023). Currently, NS generates 404.1 kg per capita waste annually with the IC&I sector contributing 60% (244.64 kg) of per capita waste, while the residential sector contributes 40% (159.46 kg) of per capita waste generation (Figure 3.1).

Waste compliance refers to the adherence and conformity of individuals, businesses or other facilities to regulations, guidelines, and standards governing the proper management, handling, and disposal of waste established by authorities to ensure that waste is properly managed to minimize environmental impact, protect public health, and promote sustainable practices (Egüez, 2021). Failure to comply with waste regulations can result in legal consequences, fines or other penalties. Achieving waste compliance is essential for promoting responsible waste management and fostering a more sustainable approach to resource use (Barchiesi et al., 2022).

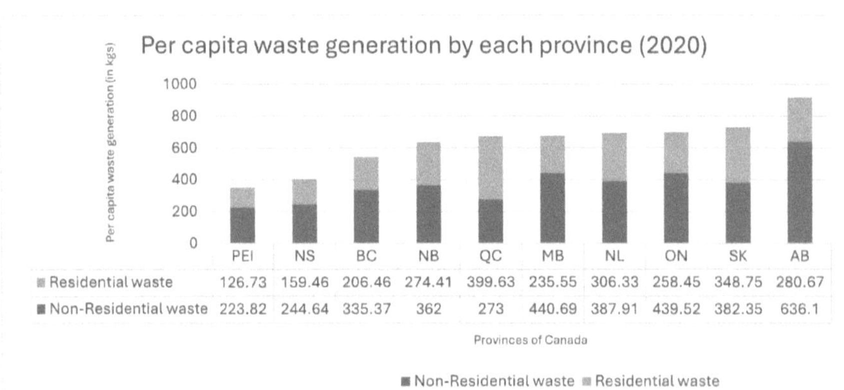

Figure 3.1 Distribution of per capita waste generation by provinces of Canada in 2020.

Note: This was created using data retrieved from Statistics Canada (2021a, 2021b).

Current scenario of IC&I waste compliance in NS

Understanding current IC&I compliance statistics is important for assessing waste management effectiveness, identifying success stories, and challenges that necessitate targeted interventions (Barchiesi et al., 2022). In NS, waste compliance checks are conducted through tipping floor inspections, visits to transfer sites, and issuing verbal or written warnings, directives and penalties (K. Bremner, personal communication, July 11, 2023). However, these approaches vary across the province and hence every region faces different challenges (K. Laskow-Pooley, personal communication, July 11, 2023).

Region 1 implemented proactive measures like distributing brochures and conducting site visits, yet still faces issues with illegal dumping and waste contamination from multi-unit buildings. Fines up to CA$697.50 have been issued but Inverness County continues to struggle with businesses exploiting residential collection services and illegal dumping practices (Divert NS, 2022). Region 2 conducted comprehensive inspections and directives to discontinue black bags have been issued. "Double tipping" is charged to noncompliant loads. Illegal dumping of construction waste remains an issue, but education and double-tipping measures have proven effective (Divert NS, 2022). Region 3 reports challenges with illegal dumping and noncompliance at Material Recovery Facilities (MRF) due to high tenant turnover. However, the Waste Wizard App has been successful in public engagement and waste compliance monitoring (Divert NS, 2022).

Region 4 in Halifax Regional Municipality (HRM) uses the 311 helpline to identify issues such as illegal dumping and litter accumulation. Systematic inspections and educational visits, along with collaboration with the RCMP, have reduced illegal dumping. The incorporation of waste room plans and cold call visits to IC&I buildings have further enhanced enforcement efficacy (Divert NS, 2022). Region 5 faces challenges with illegal dumping and tracking individuals without proper documentation. New security cameras on the tipping floor and stamped warnings on invoices have been implemented, with a penalty fee of CA$250/ton for noncompliance (Divert NS, 2022). Region 6 reports nonsorting within the IC&I sectors and C&D waste disposal without proper identification. Measures include training sessions, increased fines for illegal dumping, educational outreach, and the use of deer cameras for site monitoring. Additionally, social media platforms are used for public engagement and compliance with clear bag regulations (Divert NS, 2022). Region 7 focuses on illegal dumping at small sites and waste compliance in multi-unit dwellings. Increased calls from residents in Yarmouth indicate heightened awareness. The presence of law enforcement officers in uniform has enhanced compliance among businesses (Divert NS, 2022).

The residential sector has had more waste separation requirements compared to the IC&I sector, however, the IC&I sector tends to produce a much higher volume of waste (CCME, 2021). The objective of this chapter is to comprehensively investigate the current state of waste compliance in the IC&I sector of NS and propose evidence-based strategies for improving waste compliance and achieving sustainable waste outcomes.

Methodology

Waste compliance data within NS's IC&I sector was reviewed. Primary data sources were derived from the annual enforcement reports of 2021–22 and 2022–23, submitted by each waste management region to Divert NS. Kathryn Bremner (Director of Programs & Development, Divert NS), facilitated access to these reports including compliance statistics, barriers faced by the regions, and successful initiatives and opportunities identified during the reporting year. A detailed analysis of these reports was conducted to identify recurring themes and common barriers to IC&I compliance.

Relevant literature was also reviewed, using databases such as "Google Scholar" and "Scopus." Keywords including "multi-unit dwellings," "ICI" or "IC&I," "illegal dumping," "waste haulers," and "compliance challenges," were used to identify barriers and to develop recommendations to address specific themes identified from annual enforcement reports. Additionally, relevant waste management plans and reports from jurisdictions across Canada were examined. Previous research reports from the Divert NS database were reviewed, specifically those addressing waste management challenges in multi-unit dwellings, the C&D sector, and restaurants and fast-food chains in NS.

Insights from conversations with Kurt Laskow-Pooley and Kathryn Bremner, Divert NS, Robert Kenney, Nova Scotia Environment and Climate Change, Roschell Clarke, Cape Breton Regional Municipality, and Shannon Betts, HRM enriched the overall understanding of waste compliance dynamics.

Barriers to effective waste compliance

From illegal dumping and IC&I noncompliances at MRF to challenges unique to multi-unit dwellings, every region has its own unique set of challenges that necessitate tailored approaches to waste management (Divert NS, 2022). This section identifies some less-compliant sectors in the nonresidential category and outlines the associated major challenges to effective waste compliance in these areas.

Identifying less-compliant sectors

The multi-unit dwellings sector presents significant compliance challenges in waste management as represented in Table 3.1. Classified as commercial properties, their owners must contract private waste haulers, unlike residential sectors where this service is included in taxes. Despite bylaws mandating source separation, compliance remains low (Divert NS, 2021, 2022).

The C&D sector in NS faces significant barriers to waste compliance due to its fragmented nature (Pierre, 2014). Multiple contractors and subcontractors create challenges in monitoring compliance, leading to diffused accountability. The variety of materials generated, such as asphalt, wood, metal, concrete, and plastic, further complicates waste management. Ajayi et al. (2015) noted that C&D project owners sometimes prioritize cost and project duration over proper waste

Table 3.1 Barriers to compliance in multi-unit dwellings

Landlords/management	Unwilling to take ownership of waste generated by residents Difficult to contact if out of region or manages multiple properties Not providing correct information or bins to tenants Not understanding source separation rules
Tenants	Lack of information, diverse population Lack of proper space in units to effectively sort waste Misconceptions of management being solely responsible for waste sorting and not them Improper waste sorting facilities available in the building
Haulers	Does not provide proper information to property owners May take contaminated loads to avoid losing clients Lack of communication with regional waste management staff
Enforcement staff	Staff shortage to implement inspections and compliance Poor records of new landlords who need educational visits about regional sorting guidelines Based on certain regional bylaws, some might be unable to enforce compliance until material reaches transfer site

Source: Adapted from DeCoste (2021).

management. They propose that certain waste management practices, such as those necessitating additional man-hours, contribute to increased design and construction costs. Hence, they prefer sending out mixed wastes rather than sorted wastes to the waste management facilities. Tipping fees for mixed waste is in the range of CA$100–160/ton in most regions, hence encouraging haulers collecting mixed waste to dump it illegally (Divert NS, 2022). A growing concern in the province involves illegal dumping practices between municipalities. This arises from the desire to circumvent higher tipping fees and dispose of waste unlawfully in neighboring regions (R. Clarke, personal communication, November 23, 2023). Lack of harmonization in penalties and tipping fees across regions complicates enforcement, as different municipalities have varying rates and fines (Pierre, 2014).

The restaurant and fast-food sector also face high compliance challenges. As per the recent waste audit by Divert NS (2017), many landfills received compostable organics despite being banned. This sector is one of the biggest generators of compostable organics in the IC&I world and the fast-paced environment and poor kitchen designs further hinder proper sorting (PHA Consulting Associates, 2017). Additionally, high staff turnover rates require constant training, which may not always be feasible, especially in an industry where employees come from diverse backgrounds and may not understand English or French, which poses barriers for waste compliance. In tightly packed or busy locations, some establishments struggle to allocate sufficient space for multiple bins which leads to ineffective sorting efforts (R. Kenney, personal communication, July 11, 2023). Some restaurants rely on customer-based sorting which leads to unsorted waste as not all customers are aware of or concerned about these rules and they typically face no consequences for noncompliance (PHA Consulting Associates, 2017).

In summary, the study identified multi-residential properties, C&D sector and the restaurant and fast-food industry to be the major noncompliant sectors in IC&I waste management practices.

Other challenges to compliance

One of the biggest challenges, illegal dumping, is the deliberate and illegal act of abandoning waste in unauthorized public or private locations (Lu, 2019). It is typically done with the intention of avoiding tipping fees, reducing transportation time and costs or for the sake of convenience (Webb et al., 2006). Illegal dumping poses not only a direct nuisance, but also has other consequences (Lu, 2019). The aftermath of illegal dumping includes habitat destruction, contributing to wildlife fatalities, and soil and groundwater pollution (Webb et al., 2006). In NS, the haulers and IC&I facilities have been found to be majorly involved in illegal dumping of their unsorted waste to avoid higher tipping fees or just for the convenience of getting rid of the rejected waste (R. Clarke, personal communication, November 23, 2023). Recognizing the urgency in addressing these concerns, regional municipalities often mobilize abatement teams upon discovering instances of illegal dumping (Divert NS, 2022). This prompt response is crucial as materials may include solvents, sharps, rusted metal and batteries which contributes to severe environmental pollution, resulting in expensive cleanup operations which underscores the broader economic impact of illegal dumping on communities (Lu, 2019).

Lack of awareness among IC&I sector staff in NS poses another major barrier to waste compliance (Divert NS, 2022). High staff turnover creates a continuous challenge for waste management efforts. As employees shift roles or new personnel join the workforce, maintaining a consistent level of awareness regarding waste sorting regulations becomes difficult (RRFB, 2003). Varying sorting regulations in different places can lead to confusion among employees who may not be familiar with NS specific guidelines. Additionally, the lack of awareness extends to IC&I owners themselves, who may be uninformed about the sorting rules. This ignorance among business owners creates a “trickle-down effect”, as the absence of awareness at the managerial level makes it less likely for organizations to prioritize staff training on waste compliance (Bona & Manuel, 2023).

Private waste hauling companies play a pivotal role in the waste management ecosystem, acting as intermediaries between IC&I facilities and waste management facilities. Their responsibilities involve transporting waste generated by businesses to appropriate disposal sites, ensuring compliance with regulations and guidelines (S. Betts, personal communication, July 11, 2023). Reluctant IC&I haulers emerge as a significant barrier to waste compliance. One critical challenge arises from the absence of a province-wide formal registration requirement for waste haulers. Currently, many of the small hauling companies in the province are not registered as commercial haulers. This loophole allows some small commercial haulers to disguise themselves as residential haulers when arriving at waste transfer facilities, thereby avoiding higher tipping fees associated with commercial waste (R. Clarke, personal communication, November 23, 2023). This practice not only undermines the integrity of waste disposal fees but also hampers accurate waste categorization and compliance monitoring.

Some haulers may also be unaware of the region's waste guidelines, leading to improper waste handling and sorting at the source. This lack of awareness, coupled with inadequate communication channels between waste management regions and hauling companies worsens the problem, obstructing efforts to rectify noncompliance (R. Kenney, personal communication, July 11, 2023). There are instances where haulers do not provide the correct information and bins to IC&I facilities, resulting in noncompliance (PHA Consulting Associates, 2017). In some cases, haulers may prioritize maintaining clients by accepting contaminated loads, foregoing proper waste disposal practices to retain business. Additionally, business or private advantages may sometimes take precedence over the correct methods of waste disposal, as haulers may be motivated by financial considerations rather than environmental responsibility (Divert NS, 2022).

Analyzing factors influencing IC&I waste compliance

Tipping fees and penalties

Tipping fees and penalties play an important role in influencing waste compliance within the IC&I sector in NS. Many businesses in the IC&I sector equate and consider the perceived cost associated with sorting, storing, and transporting multiple waste bins as outweighing the cost of mixed load tipping fees. This calculation often leads to a decision to opt for mixed waste disposal, contributing to a lack of source separation and compliance (Pierre, 2014). Another crucial aspect is the wide variation in tipping fees and penalties across regions and even municipalities within the same region. For instance, tipping fees for commercial organics vary across NS, ranging from CA$51 in Colchester, CA$68.30 in Cape Breton Regional Municipality, CA$90 in HRM to CA$103 in East Hants, CA$104 in Lunenberg, CA$105 in the Valley and CA$114 in Yarmouth (Halifax Regional Council, 2023). This disparity creates a complex landscape where haulers must navigate different fee structures, making it challenging to develop a standardized approach to waste (R. Clarke, personal communication, November 23, 2023).

Enforcement visits and governmental supervision

Enforcement visits serve as visible reminders of the consequences of noncompliance, fostering a culture of adherence to IC&I waste guidelines. The element of surprise associated with enforcement visits instills a sense of fear and the unpredictability of these visits ensures that businesses remain on their toes, constantly aware of the need to adhere to waste disposal regulations, even in the absence of continuous oversight (Makofse, 2019). Through regular inspections, authorities can identify specific challenges and areas of improvement, enabling them to tailor educational workshops and outreach programs to address the unique needs of different businesses (PHA Consulting Associates, 2017).

Education programs and awareness

Effectiveness of waste compliance in the IC&I sector is closely tied to the degree of understanding among businesses, ranging from owners to employees and waste management personnel (Pierre, 2014). Raising awareness about the specific challenges faced by the IC&I sector, such as the complexities of source separation in commercial settings, can lead to increased understanding and cooperation among stakeholders (Divert NS, 2022). Education and awareness of sorting becomes necessary to address challenges associated with tenant compliance in apartment buildings. Residents need to be well informed about the proper sorting of waste to overcome barriers such as the lack of monitoring systems and active management (DeCoste, 2021).

Infrastructure and staff availability

The synergy of infrastructure and staff availability significantly influences waste compliance in the IC&I sector in NS. In the context of multi-unit dwellings, the absence of in-bin units or appropriate facilities poses a substantial hurdle, prompting tenants to engage in noncompliance with waste sorting regulations. This deficiency highlights the importance of ensuring that proper waste management infrastructure is in place within these buildings. Insufficient prioritization in the design of waste rooms in apartment buildings further adds to this issue (DeCoste, 2021). Within the construction sector, the deficiency of on-site infrastructure emerges as a major barrier to waste compliance (Pierre, 2014). Similarly, restaurants situated in downtown or busy locations face unique challenges related to space constraints for waste bins (PHA Consulting Associates, 2017). The shortage of regional waste management staff compounds challenges faced across the sector. Insufficient staffing levels can lead to fewer inspections, limited training opportunities, and limited education workshops, contributing to a decrease in overall compliance checks and undermining efforts to promote proper waste disposal practices (Divert NS, 2022).

Initiatives by waste management regions in NS to improve IC&I waste compliance

Waste management regions in NS have undertaken a series of successful initiatives to promote and enhance waste compliance in the IC&I sector. For example, these initiatives include (Divert NS, 2022):

- Law enforcement officers in uniform, which increases adherence to waste compliance regulations by businesses
- Conducting surprise inspections to the IC&I sectors

- Signage and surveillance cameras in strategic locations to reduce illegal dumping
- Regular inspections of transfer stations helped to ensure that waste disposal and handling facilities are operating in compliance with regulations
- Collaborating with local police and the RCMP to enhance waste compliance
- Partnering with education staff or professionals to raise awareness and educate businesses by providing training sessions, workshops, or educational materials
- Developing a mobile app such as Waste Wizard from Region 3 keeps everyone aware of changing bylaws and provides a platform to report waste-related issues. It has also helped the enforcement officers for an easy follow-up on reporting and prompt action.
- Issuing double tip tickets acted as a significant deterrent.

Recommendations

Promoting a transparent system with haulers

The priority of the regional waste management staff should be to enhance transparency by involving haulers and improve the overall compliance in the IC&I sector. The implementation of mandatory registration of all haulers would be a foundational step. This will not only ensure that authorized haulers are involved in waste disposal but also facilitate easy monitoring and accountability. Imposing mandatory annual reporting requirements on hauling companies is crucial. These reports should include valuable information such as the number of denied waste bags, identification of less-compliant sectors, and initiatives to enhance compliance (CCME, 2021). These reports will serve as a mechanism for both self-assessment by haulers and monitoring by regulatory bodies. To support haulers in understanding and complying with waste management regulations, the development of compliance kits is recommended. These kits, distributed to haulers upon registration, may include: a cover letter emphasizing the municipality's dedication to sustainable waste management; the region's regulatory guidelines; a guide designed to provide IC&I clients with a comprehensive resource that outlines best practices for waste segregation, handling, and disposal; a "what-goes-where" infographic, illustrating common waste streams generated by IC&I clients; a container labeling toolkit with standardized symbols and information; training resources that recommend or provide access to online resources for IC&I clients; a compliance checklist that IC&I clients can use to self-assess their compliance with regulations, and contact information for waste haulers and IC&I clients to reach out in case of issues or complaints.

A "sign-off" document can be introduced, requiring IC&I clients to agree to adhere to source separation practices when giving off waste to haulers. This document shall encourage a commitment to compliance through source separation. In cases of refusal to accept contaminated loads, a standard notice letter could be issued by the government to support haulers. This will not only acknowledge their commitment to waste management standards, but also act as a protective measure against potential disputes and business intentions. Finally, to deter repeated violations, penalties must be imposed (DeCoste, 2021).

Improving staff availability

To address the current staff shortage in NS's waste management sector, the province must prioritize allocating more staff dedicated to waste compliance (Divert NS, 2022; DeCoste, 2021). This includes recruiting and training additional personnel to enhance the capacity for inspections and regulatory enforcement. A strategic approach to increasing staff can involve the hiring of student interns. This would address the shortage and also provide valuable opportunities to gain practical experience in waste management. Alongside interns, volunteer programs can be initiated to support compliance efforts. To enhance the effectiveness of waste management regions, encouraging healthy competition among them can serve as a motivational tool (CCME, 2021; DeCoste, 2021). Recognition for regions or compliance staff members for outstanding contributions can further boost morale and encourage proactive efforts in maintaining and improving waste compliance.

Strengthening enforcement and penalties

Strengthening enforcement and penalties is also crucial for improving compliance. NS should consider standardizing tipping fees across the IC&I sector and simultaneously increase these fees for noncompliance (Clarke R., personal communication, 2023). Currently, varying tipping fee rates and penalties have been observed to have a rebound effect, leading to increased illegal dumping in regions with lower penalties (Divert NS, 2022). This affects efficient waste disposal and adds to the increased greenhouse gas emissions due to haulers traveling extra miles to save on costs. Internationally, countries showcasing some of the highest rates of waste diversion often implement high tipping fees as a disincentive against irresponsible waste disposal (Pierre, 2014). For instance, in the Netherlands, which boasts of an impressive 90% reuse or recycling rate, a combination of tipping fees, taxes, and material bans results in an average tipping fee exceeding $150/ton (Pierre, 2014.

Municipal tipping fees for general waste should be considerably higher than those for recyclables and organics. Tipping fees for recycling and organics should be set at a lower rate or even made nonexistent (CCME, 2021). Consequently, haulers would adjust their pricing systems to transfer cost advantages of source separation to IC&I owners and managers. This shift in pricing structure acts as a financial incentive for businesses to engage in responsible waste management practices. However, any restructuring of tipping fees should require careful consideration and collaboration with stakeholders to ensure a comprehensive and effective approach (CCME, 2021).

Introducing a system of double-tipping fees specifically as penalties for noncompliance can serve as a strong disincentive. Some of the regions in NS have found double-tipping fees as a great tool for ensuring compliance (Divert NS,

2022). In addition to this, holding waste haulers accountable for the proper disposal of waste is essential. In Metro Vancouver, banned materials in a load from haulers are subjected to surcharges and penalties to haulers. The Metro Vancouver bylaw also permits haulers to dispute surcharges within 30 days of issuance, yet with only 17 disputes out of 3,500 surcharges in 2018, it suggests that a relatively small number of haulers opt to challenge charges related to noncompliant loads (CCME, 2021).

Enhancing the frequency of inspections within the IC&I sector is also vital for effective enforcement (DeCoste, 2021). For facilities, especially restaurants repeatedly sending noncompliant loads, temporary closures could be implemented. This extreme measure serves as a last resort, signaling the seriousness of persistent violations and providing a strong incentive for establishments to rectify their waste management practices.

Increasing education and awareness

To enhance compliance within the IC&I sector, focus should be on education and awareness for facilities involved. Organizing community engagement events for interactive discussions is effective for fostering understanding and cooperation (PHA Consulting Associates, 2017). Adopting a "Learning by Involving" approach can also address ignored educational visits and instill proper waste management habits (Divert NS, 2022). Like fire drills, monthly waste sorting drills can transform correct methods into reflex actions. Additionally, sustainability certification programs can incentivize businesses to prioritize waste management through healthy competition (DeCoste, 2021; Pierre, 2014).

Making audit data publicly available and directly sending them to businesses promotes transparency and encourages compliance (CCME, 2021). It helps businesses prioritize waste reduction strategies. Developing multilingual brochures for waste sorting ensures that language barriers do not impede understanding (DeCoste, 2021). Given high staff turnover in the IC&I sector, focusing on manager training is essential for consistent waste management practices (PHA Consulting Associates, 2017). These sessions will equip managers to effectively communicate and enforce waste management policies within their teams.

Creating accessible mobile applications like "Waste Wizard" by Region 3 enhances engagement in waste management efforts (Divert NS, 2022). Using social media strategically to share educational content, success stories, and updates fosters a sense of community and encourages collective efforts toward a sustainable future (CCME, 2021).

Multi-unit dwellings

To enhance waste compliance in multi-unit dwellings, a comprehensive approach is essential. Firstly, establishing a centralized database with information about each apartment building, its owners, and an on-site responsible person is crucial (CCME, 2021). This facilitates effective communication, accountability, and systematic compliance monitoring. Implementing waste collection facilities on each floor and equipping units with dedicated in-unit bins for waste separation significantly enhances convenience for residents (DeCoste, 2021).

Owners must be aware of the 4Cs—Convenience, Color, Clarity, and Capacity—when planning waste rooms (DeCoste, 2021). Ensuring disposal areas are conveniently located, clearly marked with color-coded bins, and sufficient capacity fosters effective source separation. A best practice is to have a 360-liter container for every seven units (Oxford County, 2012). Addressing language barriers with multilingual brochures and posters in waste rooms can provide clear instructions on proper waste sorting (DeCoste, 2021). Including waste sorting commitments in lease agreements through a signed agreement is another effective strategy (CCME, 2021). Mandating clear bags for waste disposal enhances transparency and encourages proper sorting by making it easier for waste collectors to identify and address contamination at source (Divert NS, 2022).

Fostering social norms around responsible waste disposal through community engagement initiatives and awareness campaigns is innovative. In Vancouver, multi-unit building management encouraged residents to commit to recycling practices by signing a pledge. Participants were recognized with a distinctive sign or sticker on their doors, declaring "We recycle," creating a sense of community identity centered around environmental responsibility.

Illegal dumping of C&D waste

Illegal dumping is as a global issue and countries like United States of America, Australia and Israel have reported an observed growth in this crime. A surge in illegal dumping in Hong Kong municipality resulted in the government using big data analytics to identify illegal dumping offenders. Big data is emerging as a powerful tool that helps in the detection of urban crimes (Lu, 2019). Understanding who is dumping trash illegally is quite challenging (Divert NS, 2022), but sometimes the offenders might leave unintentional clues and exhibit hidden patterns (Lu, 2019). First, a list of indicators is developed. Then, an analytical model is created that examines these indicators to yield results. Trucks that dump illegally usually have fewer clients, spend less time at government facilities and have less waste depth in the government's records (Lu, 2019). Lu (2019) gives a detailed insight on using big data analytics to identify illegal dumping offenders. With proper record keeping, big data analytics can be used in identifying illegal dumping offenders in NS.

Additionally, to address this issue, the identification of common illegal dumping spots is important (Ichinose & Yamamoto, 2011). Following this, a strategic deployment of surveillance cameras and the installation of proper signage at identified hotspots could be done (Divert NS, 2022).

Moreover, integrating mandatory Global Positioning System tracking systems in every hauler truck will make it easy to monitor and regulate waste transportation (Lu, 2019). This would help prevent illegal dumping and also hold haulers accountable for their routes and activities.

Restaurant and fast-food sector

To enhance waste compliance in restaurants and fast-food sector, staff-based sorting should be prioritized. A study by RRFB (2003) revealed that this practice promotes high-quality source separation but also increases awareness among restaurant owners/managers about the composition of their waste stream. When staff-based sorting is unfeasible, alternative strategies should be implemented to increase customer awareness. Educational initiatives such as table tents or posters explaining proper waste sorting techniques can be strategically placed within establishments.

Ensuring the availability of proper bins with clear labels is essential both inside the restaurants and at drive-throughs. Understanding high staff turnover, the managers can play a pivotal role in extending this knowledge to their staff, creating a more informed and responsible team. To incentivize compliance, a recognition program for efficient and compliant restaurants could also be established (RRFB, 2003).

Conclusion

NS has great potential for efficient waste diversion from the IC&I sector, and this would not be possible to achieve without an equally effective compliance system. While all seven waste management regions of the province have taken successful initiatives to enforce sorting compliance, the analysis of influencing factors has revealed the significance of involving haulers and increasing transparency in waste management processes. Additionally, the study also highlights the importance of enhancing staff availability, strengthening enforcement measures and especially prioritizing standardizing tipping fees and penalties across the province. This would also help overcome barriers such as increasing illegal dumping. Lastly, in alignment with NS's ambitious goal of achieving 300 kg of waste/person/year by 2030 (Bill 57, 2021), the recommendations presented in this chapter serve as a foundation for targeted actions that can significantly contribute to the realization of this objective. By fostering collaboration, implementing robust policies, and continuously refining strategies, NS can pave the way for a more sustainable and compliant IC&I waste sector.

References

Ajayi, S. O., Oyedele, L. O., Bilal, M., Akinade, O. O., Alaka, H. A., Owolabi, H. A., & Kadiri, K. O. (2015). Waste effectiveness of the construction industry: Understanding the impediments and requisites for improvements. *Resources, Conservation and Recycling*, 102, 101–112. https://doi.org/10.1016/j.resconrec.2015.06.001.

Barchiesi, M. A., Costa, R., & Di Pillo, F. (2022). The link between the compliance with environmental legislation on separate collection and the municipal solid waste costs. *Sustainability*, 14(9), 5661. https://doi.org/10.3390/su14095661.

Bill 57, Environment Goals and Climate Change Reduction Act,1st Session, 64th General Assembly, Nova Scotia, 2021. https://nslegislature.ca/sites/default/files/legc/statutes/environmental%20goals%20and%20climate%20change%20reduction.pdf

Bona, C. A. D., & Manuel, M. S. (2023). Awareness and compliance of the students, faculty, and staff of Kalinga State University on the solid waste management program. *International Journal of English Literature and Social Sciences*, 8(3). https://doi.org/10.22161/ijels.

Canadian Council of Ministers of the Environment. (2021). *Best management practices for disposal bans, levies and incentives for end-of-life plastics*. https://ccme.ca/en/res/finaldisposalbansbmps-ensecured.pdf.

DeCoste, D. (2021). *Best practices to improve apartment building waste management, Halifax, Nova Scotia.* Divert NS. https://divertns.ca/sites/default/files/researchreportsfiles/2022-05/Best%20Practices%20to%20Improve%20Apartment%20Building%20Waste%20Management%20-%20Drew%20DeCoste%2C%20Dalhousie%20University.pdf.

Diggle, A., & Walker, T. R. (2020). Implementation of harmonized extended producer responsibility strategies to incentivize recovery of single-use plastic packaging waste in Canada. *Waste Management*, 110, 20–23.

Diggle, A., Walker, T. R., & Adams, M. (2023). Examining potential business impacts from the implementation of an extended producer responsibility program for printed paper and packaging waste in Nova Scotia, Canada. *Circular Economy*, 2(2), 100039.

DivertNS. (2017). *Waste Audit 2017.* https://divertns.ca/sites/default/files/researchreportsfiles/2021-09/WasteAudit2017.pdf.

Divert NS. (2021). *Regional waste management enforcement reports*. Unpublished report.

Divert NS. (2022). *Regional waste management enforcement reports*. Unpublished report.

Egüez, A. (2021). Compliance with the EU waste hierarchy: A matter of stringency, enforcement, and time. *Journal of Environmental Management*, *280*, 111672. https://doi.org/10.1016/j.jenvman.2020.111672

Government of Nova Scotia. (2023). *Solid waste resource management strategy*. https://novascotia.ca/nse/waste/swrmstrategy.asp#regionalcooperation.

Halifax Regional Council. (2023, May 9). *Item no. 15.1.3: Proposed amendments to Admin Order 16, respecting fees for the use of solid waste management facilities—Tip fee increases—Commercial compost*. https://cdn.halifax.ca/sites/default/files/documents/city-hall/regional-council/230509rc1513.pdf.

Ichinose, D., & Yamamoto, M. (2011). On the relationship between the provision of waste management service and illegal dumping. *Resource and Energy Economics*, 33(1), 79-93. https://doi.org/10.1016/j.reseneeco.2010.01.002.

Lu, W. (2019). Big data analytics to identify illegal construction waste dumping: A Hong Kong study. *Resources, Conservation and Recycling*, 141, 264–272.

Makofske, M. P. (2019). Inspection regimes and regulatory compliance: How important is the element of surprise? *Economics Letters*, 177, 30–34. https://doi.org/10.1016/j.econlet.2019.01.001.

Oxford County. (2012). *Multi-residential recycling: Implementing best practices*. https://thecif.ca/pdf/reports/514.4-Oxford-MRes_Final_Report.pdf.

PHA Consulting Associates. (2017). *Optimization of waste-resources management in the restaurant sector in Nova Scotia*. https://divertns.ca/sites/default/files/researchreportsfiles/2021-09/Restaurant-Sector-Report-2017.pdf.

Pierre, D. S. (2014). *Industrial symbiosis in the HRM: Upcycling waste in sustainable construction*. DivertNS. https://divertns.ca/sites/default/files/researchreportsfiles/2021-09/Final%20Report%20%28draft%29%20-%20Davin%20St%20Pierre_compressed.pdf.

RRFB. (2003). *Quick-service restaurant waste reduction study final report*. https://divertns.ca/sites/default/files/researchreportsfiles/2021-09/RRFB-2003_Restaurant_Study.pdf.

Statistics Canada. (2021a). *Table 17-10-0009-01 Population estimates, quarterly*. https://doi.org/10.25318/1710000901-eng.

Statistics Canada. (2021b). *Table 38-10-0032-01 Disposal of waste, by source*. https://doi.org/10.25318/3810003201-eng.

Statistics Canada. (2024). *2023 Annual Waste Management Survey*. www.statcan.gc.ca/en/statistical-programs/instrument/2009_Q1_V13.

Webb, B., Marshall, B., Czarnomski, S., & Tilley, N. (2006). *Fly-tipping: Causes, incentives and solutions*. https://rb.gy/px4bur

4 Bridging the gap

Communicating sustainability in waste management to diverse audiences

Calvin Lakhan

Introduction

The field of waste management has undergone a significant transformation in recent decades, shifting from a focus on mere disposal to a comprehensive approach that emphasizes sustainability, resource recovery, and environmental stewardship. As this evolution continues, the language we use to discuss these concepts plays a crucial role in shaping public understanding and participation in sustainable waste management practices. However, communicating these often complex ideas to diverse audiences presents unique challenges, particularly when engaging with ethnic communities who may be unfamiliar with sustainability terminology (Atkins & Moy, 2021).

This chapter explores the intricate relationship between language, culture, and sustainability in the context of waste management. We will examine the barriers that hinder effective communication of sustainability concepts and propose innovative solutions to overcome these obstacles. Our focus will be on developing strategies that bridge the gap between technical expertise and public understanding, with a particular emphasis on engaging ethnic communities in the dialogue of sustainable waste management.

The importance of this topic cannot be overstated. As urbanization increases and communities become more diverse, the success of sustainable waste management initiatives increasingly depends on the active participation of all community members. Without clear, culturally sensitive communication, even the most well-designed waste management programs risk failure. Moreover, the exclusion of certain communities from this vital conversation raises serious questions of environmental justice and equity (Bullard & Wright, 2018).

Throughout this chapter, we will draw upon theoretical frameworks and practical examples to illustrate both the challenges and potential solutions in communicating sustainability. We will explore how language barriers, cultural differences, varying levels of environmental literacy, and socioeconomic factors all contribute to the complexity of this issue. By the end of this chapter, readers will have a comprehensive understanding of the role of language in sustainable waste management

DOI: 10.4324/9781003534129-6

and be equipped with strategies to effectively communicate these concepts across diverse cultural contexts.

As professionals in the field of waste management, it is essential to develop not only technical expertise but also the ability to communicate complex ideas to a wide audience. The skills and insights gained from this chapter will be invaluable in creating inclusive, effective, and truly sustainable waste management systems that serve all members of our increasingly diverse communities.

The importance of inclusive communication in waste management

The role of public participation in sustainable waste management

Public participation is not merely a desirable component of waste management; it is an essential driver of sustainable practices. When community members are actively engaged in waste management processes, the outcomes are significantly more effective and long-lasting (Reed et al., 2018).

Public participation leads to more informed decision-making. Local communities possess invaluable knowledge about their specific waste generation patterns, cultural practices, and environmental concerns. By incorporating this knowledge into waste management strategies, authorities can develop more targeted and effective solutions. For instance, a community with a high proportion of multigenerational households might require different waste collection methods compared to an area dominated by single-occupancy apartments.

Active public involvement fosters a sense of ownership and responsibility. When individuals feel that their voices are heard and their input is valued, they are more likely to comply with waste management regulations and actively participate in initiatives such as recycling or composting programs (Beierle & Cayford, 2015). This sense of ownership can transform passive recipients of waste management services into proactive partners in sustainability efforts.

Public participation can also lead to innovation. Community members often devise creative solutions to local waste problems that professionals might overlook. In some communities, residents have initiated tool-sharing programs to reduce the purchase of rarely used items, effectively reducing waste at the source (Wilson et al., 2015).

However, achieving meaningful public participation requires clear, accessible communication. Technical jargon and complex concepts must be translated into language that resonates with diverse community members, enabling them to engage fully in the waste management discourse (Chess & Purcell, 2019).

The diversity of stakeholders in waste management systems

Waste management systems involve a wide array of stakeholders, each with unique perspectives, needs, and roles. Recognizing and effectively communicating with this diverse group is crucial for the success of any sustainable waste management initiative (Scheinberg et al., 2016).

Key stakeholders typically include:

1 Waste generators: Including residents and businesses, their actions significantly impact the effectiveness of waste management systems (Zorpas & Lasaridi, 2013).
2 Government agencies: At local, regional, and national levels, they set policies, enforce regulations, and often oversee waste management operations.
3 Waste management professionals: Including collectors, sorters, recyclers, and landfill operators, they manage waste on a day-to-day basis.
4 Environmental groups: They advocate for sustainable practices and often serve as watchdogs for environmental compliance (Fisher et al., 2019).
5 Educational institutions: They play a vital role in educating future generations about sustainable waste management.
6 Informal waste sector workers: In many parts of the world, informal waste pickers and recyclers are integral to waste management systems (Aparcana, 2017).
7 Technology providers: They develop new waste management technologies, from advanced recycling processes to smart bin systems (Esmaeilian et al., 2018).

Each of these stakeholder groups may have different levels of technical knowledge, priorities, and cultural backgrounds. Effective communication must be tailored to address the specific needs and perspectives of each group while also fostering a shared understanding of sustainable waste management goals.

The consequences of excluding communities from the sustainability dialogue

The exclusion of any community from the sustainability dialogue in waste management can have far-reaching and detrimental consequences, both for the excluded groups and for the overall effectiveness of waste management systems.

1 Environmental injustice: Historically, marginalized communities have often borne a disproportionate burden of waste management infrastructure, such as landfills or incinerators, while being excluded from decision-making processes (Bullard & Wright, 2018). This exclusion perpetuates cycles of inequality and can result in long-term health and environmental impacts on these communities.
2 Ineffective systems: When certain groups are left out of the sustainability conversation, it can lead to the development of waste management systems that are not suited to their needs or cultural practices. This mismatch can result in low participation rates, improper waste disposal, and ultimately, the failure of well-intentioned initiatives (Agyeman et al., 2016).
3 Missed opportunities: Exclusion represents a missed opportunity to tap into diverse knowledge systems and innovative solutions. Different cultural traditions often encompass valuable sustainability practices that could inform and enhance modern waste management approaches (Meinzen-Dick, 2017).

4 Lack of trust: The exclusion of communities can lead to a lack of trust in waste management authorities. When people feel their concerns are not heard or their needs are not considered, they may become resistant to new initiatives, even if these initiatives could benefit their community (Petts & Leach, 2017).
5 Global repercussions: In our interconnected world, the exclusion of any group from sustainable practices can have global repercussions. Waste, particularly plastic waste, does not respect borders. The failure to engage all communities in sustainable waste management practices can contribute to widespread environmental degradation that affects us all (Jambeck et al., 2015).

Inclusive communication in waste management is not just a matter of fairness or political correctness. It is a practical necessity for developing effective, equitable, and truly sustainable waste management systems. By ensuring that all voices are heard and all communities are engaged, we can create more resilient, innovative, and successful approaches to managing our waste in an environmentally responsible manner.

Challenges in communicating sustainability concepts

Language barriers

Language barriers represent a fundamental challenge in communicating sustainability concepts, especially in multicultural societies. The complexity of this issue extends beyond mere translation and encompasses several interrelated aspects:

1 Technical jargon and complex terminology: Terms like "circular economy," "life cycle assessment," or "anaerobic digestion" are commonplace in professional and academic circles but may be incomprehensible to the general public (Habib et al., 2020). This specialized vocabulary, while precise and necessary for experts, can create a significant barrier to understanding for those outside the field. The challenge lies in conveying these complex ideas without losing their essential meaning or oversimplifying to the point of inaccuracy.
2 Limited availability of multilingual resources: In many regions, information about sustainable waste management practices is primarily available in the dominant language, leaving nonnative speakers at a disadvantage (Stewart & Niero, 2018). This scarcity of multilingual materials can lead to the exclusion of linguistic minorities from important conversations about sustainability and waste management in their communities.
3 Idiomatic expressions and cultural references: Sustainability concepts are often explained using metaphors or analogies that may not translate well across languages or cultures. For instance, the concept of a "carbon footprint" might be challenging to convey in cultures where this metaphorical use of "footprint" doesn't exist (Holmes, 2019). Similarly, references to specific cultural practices or local environmental issues used to illustrate sustainability concepts may not resonate with all audiences.

When key terminologies cannot be translated, it's important to consider using the original term followed by a brief explanation or relatable analogy in the target language.

Cultural differences

Cultural differences play a significant role in how sustainability concepts are perceived, understood, and acted upon. These differences can create substantial barriers to effective communication and implementation of sustainable waste management practices:

1. Varying perceptions of waste and cleanliness: In some cultures, the concept of waste is closely tied to notions of impurity or uncleanliness, which can make discussions about waste management challenging (Schultz et al., 2020). Other cultures may have traditions of reuse and recycling that predate modern sustainability concepts but are not recognized as such. Understanding and respecting these cultural perceptions is crucial for effective communication.
2. Different priorities and values related to environmental issues: Some cultures may place a higher value on immediate economic concerns over long-term environmental sustainability. In contrast, others might have deep-rooted traditions of environmental stewardship that align closely with modern sustainability concepts (Inglehart & Welzel, 2015). Communicators must navigate these varying priorities to make sustainability messages relevant and compelling to diverse audiences.
3. Cultural taboos or sensitivities: In some cultures, discussing certain types of waste (such as menstrual waste or food waste) may be considered inappropriate or offensive. These cultural sensitivities must be carefully considered when developing communication strategies to avoid alienating or offending the target audience (Douglas, 2013).

Educational barriers

Educational barriers present a significant challenge in communicating sustainability concepts, as they can limit individuals' ability to engage with and understand complex environmental issues:

1. Varying levels of environmental literacy: Some individuals may have a deep understanding of environmental issues, while others may have little to no formal education on the subject (McBeth et al., 2019). This disparity necessitates communication strategies that can effectively reach audiences with diverse levels of prior knowledge.
2. Limited access to sustainability education: Many individuals may not have had the opportunity to learn about environmental issues in formal educational settings. This lack of foundational knowledge can make it challenging to engage with more complex sustainability concepts related to waste management (UNESCO, 2019).

3 Misconceptions about waste management practices: Common myths, such as the belief that all plastic is recyclable or that landfills are a sustainable solution for waste disposal, can be deeply ingrained and difficult to dispel (Geyer et al., 2017). Overcoming these misconceptions requires not only providing accurate information but also addressing the underlying beliefs and assumptions that support them.

Socioeconomic factors

Socioeconomic factors play a crucial role in how sustainability concepts are communicated and received, often creating significant barriers to effective engagement with diverse communities:

1 Digital divide and disparities in access to information: In an age where much sustainability information is disseminated online, communities with limited internet access or low digital literacy may be excluded from important conversations and resources (Robinson et al., 2015). This digital divide can reinforce existing inequalities and prevent certain groups from fully participating in sustainability initiatives.
2 Time and resource constraints: For many people, especially in economically disadvantaged communities, immediate concerns such as meeting basic needs may take precedence over long-term environmental considerations. The challenge lies in making sustainability relevant and accessible to those who may not have the luxury of time or resources to prioritize environmental issues (Martinez-Alier et al., 2016).
3 Competing priorities in disadvantaged communities: In areas facing urgent social or economic challenges, environmental concerns may be perceived as less pressing. Communicators must find ways to connect sustainability in waste management to other community priorities, such as public health, economic opportunity, or quality of life improvements (Agyeman, 2013).
4 Practical barriers to implementation: Socioeconomic factors can influence individuals' ability to act on sustainability information. For instance, participating in recycling programs may be more challenging for those living in multi-family housing without adequate recycling infrastructure, or adopting zero-waste practices may be financially unfeasible for low-income families. Effective communication must not only convey information but also acknowledge these practical barriers and provide realistic, accessible solutions (Schröder et al., 2019).

The challenges in communicating sustainability concepts in waste management are multifaceted and interconnected. Language barriers, cultural differences, educational disparities, and socioeconomic factors all contribute to the complexity of this task. Addressing these challenges requires a nuanced, interdisciplinary approach that considers the diverse needs, perspectives, and circumstances of different communities. By recognizing and strategically addressing these barriers, we can develop more effective, inclusive, and impactful communication strategies for sustainable waste management.

Strategies for effective communication

Linguistic adaptations

Linguistic adaptations are fundamental to ensuring that sustainability messages reach and resonate with diverse audiences. This strategy involves more than mere translation; it requires a nuanced approach to language use that considers the linguistic diversity of the target community (Krauss & Chiu, 2018).

1 Developing multilingual communication materials: Create resources in the primary languages spoken in the community. Employ professional translators familiar with both the technical aspects of waste management and the cultural nuances of the target language. This ensures that the translated materials convey not just the words, but also the intended meaning and significance of the sustainability concepts (Byrne, 2016).
2 Using plain language and avoiding jargon: Replace technical terms with simpler, more accessible language or, when unavoidable, explain them clearly in context. For example, instead of using the term "source separation," materials could refer to "sorting your waste at home" (Willerton, 2015).
3 Employing visual aids and infographics: Use visual representations to help overcome language barriers and improve comprehension. For instance, create easily recognizable and culturally appropriate visual guides for waste sorting (Lazard & Atkinson, 2015).

Cultural competence

Cultural competence in communication involves understanding and respecting the cultural norms, values, and practices of diverse communities. This understanding should inform every aspect of the communication strategy (Betancourt et al., 2016).

1 Conducting cultural research: Explore the target community's attitudes toward waste, environmental issues, and sustainability. Investigate any cultural taboos or sensitivities related to waste management. For example, in some cultures, handling waste is considered unclean or degrading, which could impact participation in recycling programs (Douglas, 2013).
2 Collaborating with community leaders and cultural liaisons: Work with respected members of various ethnic and cultural communities who can act as bridges between their communities and waste management authorities. They can provide valuable insights into community perspectives and help tailor communication strategies in a culturally appropriate manner (Israel et al., 2017).
3 Adapting messages to align with cultural values and practices: Frame sustainability messages in terms that resonate with the community's cultural values.

For instance, in communities with strong religious traditions, sustainability messages could be framed in terms of environmental stewardship as a religious duty. In cultures with a strong emphasis on family, messages could highlight how sustainable practices benefit future generations (Schwartz, 2015).

Community engagement

Community engagement is a powerful strategy for effective communication of sustainability concepts. It involves actively involving community members in the development and implementation of waste management initiatives (Reed et al., 2018).

1 Organizing participatory workshops and focus groups: Allow community members to voice their concerns, share their ideas, and contribute to the development of waste management strategies. These forums can provide valuable insights into the community's needs and preferences, helping to shape more effective and acceptable initiatives (Bryson et al., 2017).
2 Utilizing community-based social marketing techniques: Identify barriers to sustainable behaviors within the community and develop targeted strategies to overcome these barriers. For example, if lack of knowledge about proper recycling practices is identified as a key barrier, the strategy might involve door-to-door education campaigns or community recycling demonstrations (McKenzie-Mohr, 2011).
3 Empowering local ambassadors for sustainability: Choose ambassadors from within the community to act as role models and educators, sharing their knowledge and enthusiasm with their peers. This peer-to-peer approach can be particularly effective in communities where there may be distrust of outside authorities (Rogers, 2015).
4 Implementing behavioral change interventions: Utilize strategies such as social proof, commitment devices, and nudges to encourage sustainable waste management behaviors (Yoeli et al., 2017).

Educational initiatives

Educational initiatives play a crucial role in building a foundation of knowledge about sustainability and waste management within diverse communities.

1 Developing culturally relevant environmental education programs: Create programs that incorporate local environmental issues and cultural perspectives on nature and waste. For example, in coastal communities, education programs might focus on the impact of waste on marine ecosystems, while in agricultural areas, the focus might be on composting and soil health (Monroe et al., 2017).
2 Partnering with schools and community organizations: Integrate waste management and sustainability concepts into school curricula, reaching young people who can then become advocates within their families. Collaborate with

community organizations to host workshops and events, providing platforms for ongoing education and discussion (Ardoin et al., 2018).

3 Creating hands-on learning experiences: Organize field trips to recycling facilities, set up community composting projects, or run upcycling workshops. These experiential learning opportunities allow participants to see the impact of sustainable practices firsthand (Kolb & Kolb, 2017).
4 Leveraging social media as educational tools: Utilize popular social media platforms to disseminate information, create online communities, and engage younger generations in sustainability discussions (Dookhoo, 2019).
5 Involving universities and research institutions in knowledge mobilization: Collaborate with academic institutions to translate research findings into practical, community-friendly information and initiatives (Trencher et al., 2014).
6 Engaging informal waste pickers in education and awareness creation: Recognize and involve informal waste sector workers as valuable educators and ambassadors for sustainable waste management practices (Gutberlet, 2015).
7 Identifying and supporting waste management "community champions": Empower local individuals who are passionate about sustainability to lead educational efforts and drive change within their communities (Sharp, 2016).

Leveraging technology

Technology offers powerful tools for communicating sustainability concepts effectively across diverse communities.

1 Developing multilingual mobile applications: Create user-friendly apps that provide easily accessible information about waste management practices, including waste sorting guides, collection schedules, and locations of recycling facilities (Lim et al., 2019).
2 Using social media platforms popular within specific communities: Identify and utilize the most popular social media channels in target communities to disseminate information, foster online communities, and encourage ongoing engagement with sustainability topics (Zhou & Wang, 2014).
3 Creating virtual reality experiences: Develop immersive virtual reality tours of waste management facilities or interactive simulations of sustainable practices to make complex concepts more understandable and engaging (Markowitz et al., 2018).
4 Implementing smart waste management systems: Introduce technologies like IoT-enabled smart bins and route optimization systems, which can improve waste management efficiency and provide real-time data for public education and engagement (Esmaeilian et al., 2018).

By implementing these diverse strategies, waste management initiatives can better engage diverse communities, leading to more successful and inclusive sustainability outcomes. It's important to note that the effectiveness of these approaches may vary depending on the specific context and community, and they should be adapted and combined as needed to achieve the best results.

Measuring success and continuous improvement

Developing metrics for communication effectiveness

Developing meaningful metrics is essential for assessing the impact of sustainability communication efforts and identifying areas for improvement. These metrics should go beyond simple quantitative measures to capture the nuanced effects of communication strategies on diverse communities.

When planning effective communication strategies, it's crucial to clearly define what constitutes "success" alongside its indicators (both quantitative and qualitative). This provides a pathway to measuring "success" and guides the development of appropriate metrics.

1. Reach metrics: Track not just the number of people exposed to sustainability messages, but also the demographic breakdown of this audience. For instance, measure the percentage of non–English-speaking households reached, or the representation of different age groups in engagement efforts (Dale et al., 2019). These data can help identify any gaps in outreach and inform targeted communication strategies.
2. Engagement metrics: Measure attendance at sustainability workshops, participation rates in recycling programs, or interactions with online resources (Moser & Pike, 2015). More sophisticated engagement metrics might track the depth of engagement, such as the percentage of participants who move from awareness to action, or the number of community members who become sustainability ambassadors.
3. Behavior change metrics: Conduct waste audits to measure changes in recycling rates or contamination levels, or survey households about their adoption of waste reduction practices. Longitudinal studies that track behavior changes over time can provide particularly valuable insights.
4. Comprehension metrics: Implement pre- and post-communication assessments to measure improvements in understanding of key waste management principles (McKenzie-Mohr, 2011). For multilingual communities, conduct comprehension metrics in multiple languages to ensure accurate assessment across all groups.
5. Behavioral change intervention metrics: Assess the effectiveness of specific interventions by measuring changes in targeted behaviors before and after implementation (Yoeli et al., 2017).

Gathering feedback from diverse communities

Gathering robust, representative feedback from diverse communities is crucial for understanding the effectiveness of communication strategies and identifying areas for improvement. This process should be ongoing and should employ a variety of methods to ensure comprehensive input.

1. Multilingual surveys: Design surveys in multiple languages and make them accessible through various channels (online, paper, phone) to accommodate different preferences and access levels. Use simple and clear language, avoiding jargon or complex terms that might be misunderstood (Dillman et al., 2014).
2. Focus groups: Conduct focus groups in community languages and include representation from various demographic groups. Employ skilled facilitators who understand the cultural nuances of the community to ensure all participants feel comfortable sharing their honest opinions (Krueger & Casey, 2014).
3. Community forums or town hall meetings: Organize events that allow for open discussion and can reveal community concerns or ideas that might not emerge through more structured feedback methods (Nabatchi & Leighninger, 2015). Hold these forums at varying times and locations, with provisions for childcare and transportation where possible.
4. Digital platforms: Leverage social media polls, online discussion forums, or dedicated feedback apps. Remember that not all community members may have equal access to or comfort with digital tools, so these should complement, not replace, other feedback methods (Bonson et al., 2015).
5. Feedback mechanism evaluation: Regularly assess the feedback gathering process, tools, and models to ensure that they are truly inclusive and capturing input from all segments of the community (Patton, 2017).

Adapting strategies based on ongoing evaluation

The true value of measuring success and gathering feedback lies in using these insights to continuously improve communication strategies. This adaptive approach ensures that sustainability communication remains relevant, effective, and responsive to community needs.

1. Regular review sessions: Hold sessions to analyze metrics and feedback data. Involve a diverse team, including waste management professionals, communication experts, and community representatives. Identify trends, successes, and areas for improvement in current communication strategies (Schein, 2016).
2. Strategy refinement: Based on these reviews, refine and update communication strategies. This might involve adjusting the language used in materials, changing the channels of communication, or modifying the content of educational programs. For instance, if feedback indicates that certain concepts are consistently misunderstood, allocate more resources to explaining these ideas, perhaps through new visual aids or hands-on demonstrations (Kotter, 2012).
3. Pilot testing: Before full-scale implementation, conduct pilot tests of new or modified strategies. This allows for fine-tuning based on initial responses and can help avoid large-scale failures. Conduct pilot tests with diverse groups

within the community to ensure the strategies work across different demographics (Bryson, 2018).

4 Staying informed: Keep abreast of broader trends in sustainability communication and waste management. Attend conferences, participate in professional networks, or collaborate with academic institutions. Insights from these sources can inspire new approaches or help anticipate future challenges (Wals, 2019).
5 Transparency in adaptation: Make the process of adaptation transparent and inclusive. Inform communities about how their feedback is being used to shape communication strategies. This not only demonstrates responsiveness but also encourages ongoing community engagement in the process (Fung, 2015).

Conclusion

The language of sustainability in waste management is not merely a collection of technical terms and concepts; it is a powerful tool for engaging diverse communities in the crucial task of creating a more sustainable future. By recognizing the power of inclusive communication, addressing challenges head-on, and implementing innovative solutions, we can create a shared language of sustainability that transcends cultural and linguistic boundaries. This shared language has the potential to unite diverse communities in the common goal of creating a more sustainable, equitable, and waste-conscious world.

As professionals in the field of waste management, we are poised to play a crucial role in this important work. By applying the principles and strategies outlined in this chapter, we can help ensure that the language of sustainability is accessible, meaningful, and motivating to all members of the communities we serve. In doing so, we will not only advance the goals of sustainable waste management but also contribute to building more inclusive, environmentally conscious societies.

The journey toward a sustainable future is a collective one, and effective communication is one of the vehicles that will take us there. By mastering the art of communicating sustainability across diverse communities, we can turn the challenge of waste management into an opportunity for global cooperation, innovation, and positive change.

References

Agyeman, J. (2013). *Introducing just sustainabilities: Policy, planning, and practice.* Zed Books.

Agyeman, J., Schlosberg, D., Craven, L., & Matthews, C. (2016). Trends and directions in environmental justice: From inequity to everyday life, community, and just sustainabilities. *Annual Review of Environment and Resources*, 41, 321–340.

Aparcana, S. (2017). Approaches to formalization of the informal waste sector into municipal solid waste management systems in low- and middle-income countries: Review of barriers and success factors. *Waste Management*, 61, 593–607.

Ardoin, N. M., Bowers, A. W., & Gaillard, E. (2018). Environmental education outcomes for conservation: A systematic review. *Biological Conservation*, 241, 108224.

Atkins, J. F., & Moy, F. (2021). Sustainable waste management: International perspectives. In *Handbook of sustainability science and research* (pp. 159–176). Springer.
Beierle, T. C., & Cayford, J. (2015). *Democracy in practice: Public participation in environmental decisions*. Routledge.
Betancourt, J. R., Green, A. R., & Carrillo, J. E. (2016). *Cross-cultural care and communication*. UpToDate.
Bonson, E., Torres, L., Royo, S., & Flores, F. (2015). Local e-government 2.0: Social media and corporate transparency in municipalities. *Government Information Quarterly*, 32(2), 154–161.
Bryson, J. M. (2018). *Strategic planning for public and nonprofit organizations: A guide to strengthening and sustaining organizational achievement*. John Wiley & Sons.
Bryson, J. M., Quick, K. S., Slotterback, C. S., & Crosby, B. C. (2017). Designing public participation processes. *Public Administration Review*, 73(1), 23–34.
Bullard, R. D., & Wright, B. (2018). *The wrong complexion for protection: How the government response to disaster endangers African American communities*. NYU Press.
Byrne, J. (2016). Improving the peer review of narrative literature reviews. *Research Integrity and Peer Review*, 1(1), 12.
Chess, C., & Purcell, K. (2019). Public participation and the environment: Do we know what works? *Environmental Science & Technology*, 33(16), 2685–2692.
Dale, A., Ling, C., & Newman, L. (2019). Community vitality: The role of community-level resilience adaptation and innovation in sustainable development. *Sustainability*, 2(1), 215–231.
Dillman, D. A., Smyth, J. D., & Christian, L. M. (2014). *Internet, phone, mail, and mixed-mode surveys: The tailored design method*. John Wiley & Sons.
Dookhoo, S. (2019). Social media as a tool for increased environmental awareness. *Environmental Scientist*, 28(1), 52–55.
Douglas, M. (2013). *Purity and danger: An analysis of concepts of pollution and taboo*. Routledge.
Esmaeilian, B., Wang, B., Lewis, K., Duarte, F., Ratti, C., & Behdad, S. (2018). The future of waste management in smart and sustainable cities: A review and concept paper. *Waste Management*, 81, 177–195.
Fisher, D. R., Campbell, L. K., & Svendsen, E. S. (2019). The organisational structure of urban environmental stewardship. *Environmental Politics*, 21(1), 26–48.
Fung, A. (2015). Putting the public back into governance: The challenges of citizen participation and its future. *Public Administration Review*, 75(4), 513–522.
Geyer, R., Jambeck, J. R., & Law, K. L. (2017). Production, use, and fate of all plastics ever made. *Science Advances*, 3(7), e1700782.
Gutberlet, J. (2015). Cooperative urban mining in Brazil: Collective practices in selective household waste collection and recycling. *Waste Management*, 45, 22–31.
Habib, K., Schmidt, J. H., & Christensen, P. (2020). A historical perspective of global warming potential from municipal solid waste management. *Waste Management*, 119, 148–159.
Holmes, J. (2019). *An introduction to sociolinguistics*. Routledge.
Inglehart, R., & Welzel, C. (2015). *Modernization, cultural change, and democracy: The human development sequence*. Cambridge University Press.
Israel, B. A., Eng, E., Schulz, A. J., & Parker, E. A. (Eds.). (2017). *Methods for community-based participatory research for health*. John Wiley & Sons.

Jambeck, J. R., Geyer, R., Wilcox, C., Siegler, T. R., Perryman, M., Andrady, A., Narayan, R., & Law, K. L. (2015). Plastic waste inputs from land into the ocean. *Science*, 347(6223), 768–771.

Kolb, A. Y., & Kolb, D. A. (2017). Experiential learning theory as a guide for experiential educators in higher education. *Experiential Learning & Teaching in Higher Education*, 1(1), 7–44.

Kotter, J. P. (2012). *Leading change*. Harvard Business Review Press.

Krauss, R. M., & Chiu, C. Y. (2018). Language and social behavior. In D. T. Gilbert, S. T. Fiske, & G. Lindzey (Eds.), *The handbook of social psychology* (pp. 41–88). McGraw-Hill.

Krueger, R. A., & Casey, M. A. (2014). *Focus groups: A practical guide for applied research*. Sage publications.

Lazard, A., & Atkinson, L. (2015). Putting environmental infographics center stage: The role of visuals at the elaboration likelihood model's critical point of persuasion. *Science Communication*, 37(1), 6–33.

Lim, M. K., Mak, H. Y., & Rong, Y. (2019). Toward mass adoption of electric vehicles: Impact of the range and resale anxieties. *Manufacturing & Service Operations Management*, 17(1), 101–119.

Markowitz, D. M., Laha, R., Perone, B. P., Pea, R. D., & Bailenson, J. N. (2018). Immersive virtual reality field trips facilitate learning about climate change. *Frontiers in Psychology*, 9, 2364.

Martinez-Alier, J., Temper, L., Del Bene, D., & Scheidel, A. (2016). Is there a global environmental justice movement? *Journal of Peasant Studies*, 43(3), 731–755.

McBeth, W., Hungerford, H., Marcinkowski, T., Volk, T., & Meyers, R. (2019). *National environmental literacy assessment, phase two: Measuring the effectiveness of North American environmental education programs with respect to the parameters of environmental literacy. Final research report*. NOAA.

McKenzie-Mohr, D. (2011). *Fostering sustainable behavior: An introduction to community-based social marketing*. New Society Publishers.

Meinzen-Dick, R. (2017). Beyond panaceas in water institutions. *Proceedings of the National Academy of Sciences*, 104(39), 15200–15205.

Monroe, M. C., Plate, R. R., Oxarart, A., Bowers, A., & Chaves, W. A. (2017). Identifying effective climate change education strategies: A systematic review of the research. *Environmental Education Research*, 25(6), 791–812.

Moser, S. C., & Pike, C. (2015). Community engagement on adaptation: Meeting a growing capacity need. *Urban Climate*, 14, 111–115.

Nabatchi, T., & Leighninger, M. (2015). *Public participation for 21st century democracy*. John Wiley & Sons.

Patton, M. Q. (2017). *Principles-focused evaluation: The guide*. Guilford.

Petts, J., & Leach, B. (2017). *Evaluating methods for public participation: Literature review*. Environment Agency.

Reed, M. S., Vella, S., Challies, E., de Vente, J., Frewer, L., Hohenwallner-Ries, D., … & van Delden, H. (2018). A theory of participation: what makes stakeholder and public engagement in environmental management work? *Restoration Ecology*, 26, S7–S17.

Robinson, L., Cotten, S. R., Ono, H., Quan-Haase, A., Mesch, G., Chen, W., … & Stern, M. J. (2015). Digital inequalities and why they matter. *Information, Communication & Society*, 18(5), 569–582.

Rogers, E. M. (2015). *Diffusion of innovations*. Simon and Schuster.

Schein, E. H. (2016). *Organizational culture and leadership*. John Wiley & Sons.

Scheinberg, A., Nesić, J., Savain, R., Luppi, P., Syet, S., Ignatieva, A., & Pop, J. (2016). From collision to collaboration: Integrating informal recyclers and re-use operators in Europe: A review. *Waste Management & Research*, 34(9), 820–839.
Schröder, P., Lemille, A., & Desmond, P. (2019). Making the circular economy work for human development. *Resources, Conservation and Recycling*, 156, 104686.
Schultz, P. W., Nolan, J. M., Cialdini, R. B., Goldstein, N. J., & Griskevicius, V. (2020). The constructive, destructive, and reconstructive power of social norms: Reprise. *Perspectives on Psychological Science*, 15(5), 1190–1201.
Schwartz, S. H. (2015). Basic individual values: Sources and consequences. In D. Sander & T. Brosch (Eds.), *Handbook of value: Perspectives from economics, neuroscience, philosophy, psychology and sociology* (pp. 63–84). Oxford University Press.
Sharp, L. (2016). Understanding social learning and its contribution to sustainability education. In *Engaging stakeholders in education for sustainable development at university level* (pp. 55–71). Springer.
Stewart, R., & Niero, M. (2018). Circular economy in corporate sustainability strategies: A review of corporate sustainability reports in the fast-moving consumer goods sector. *Business Strategy and the Environment*, 27(7), 1005–1022.
Trencher, G., Bai, X., Evans, J., McCormick, K., & Yarime, M. (2014). University partnerships for co-designing and co-producing urban sustainability. *Global Environmental Change*, 28, 153–165.
UNESCO. (2019). *Educational content up close: Examining the learning dimensions of education for sustainable development and global citizenship education*. UNESCO.
Wals, A. E. (2019). Sustainability-oriented ecologies of learning. In *Learning ecologies: Sightings, possibilities, and emerging practices* (pp. 61–78). Springer.
Willerton, R. (2015). *Plain language and ethical action: A dialogic approach to technical content in the 21st century*. Routledge.
Wilson, D. C., Rodic, L., Modak, P., Soos, R., Carpintero, A., Velis, K., … & Simonett, O. (2015). *Global waste management outlook*. United Nations Environment Programme.
Yoeli, E., Budescu, D. V., Carrico, A. R., Delmas, M. A., DeShazo, J. R., Ferraro, P. J., … & Weber, E. U. (2017). Behavioral science tools to strengthen energy & environmental policy. *Behavioral Science & Policy*, 3(1), 68–79.
Zhou, X., & Wang, X. (2014). Ecologically friendly energy efficient scheduling strategy for cyber-enabled flexible job shop. *Journal of Robotics and Computer-Integrated Manufacturing*, 30(5), 503–510.
Zorpas, A. A., & Lasaridi, K. (2013). Measuring waste prevention. *Waste Management*, 33(5), 1047–1056.

Part 2

Waste management in northern, remote, rural and Indigenous communities

5 Waste stewardship in northern, remote, rural and First Nations communities in Saskatchewan

Challenges and opportunities

Amy Richter, Nima Karimi and Kelvin Tsun Wai Ng

Introduction

Extended Producer Responsibility (EPR) is a policy approach which requires that the producer be physically and/or financially responsible for managing waste that they put on the market (Wang et al., 2023). Saskatchewan uses EPR to manage a variety of wastes, particularly those that are persistent or tough to manage like tires, household hazardous waste, used oil and others (Government of Saskatchewan, 2020). With the exception household packaging and paper, which is generally collected curbside, other regulated wastes in Saskatchewan (i.e., those captured through EPR programs) require residents to drop off their waste at designated locations (Saskatchewan Waste Reduction Council (SWRC, 2024). SWRC provides a waste reduction hub that allows residents to find where they can recycle different products in their communities (SWRC, 2024). Options for recycling in northern Saskatchewan are limited compared to the southern half of the province (SWRC, 2024), likely due to differences in population density and remoteness. A better understanding of the interplay between barriers to recycling in northern, remote and First Nations communities, along with an analysis of the effects of government policies may lead to a data-driven solution to improve recycling efficiency and effectiveness in those areas.

Waste management in isolated regions

Waste management is a challenge in small, isolated regions (Sebastian & Louis, 2022), characteristic of much of northern Saskatchewan. Studies on Saskatchewan, northern Ontario and the Northwest Territories found that a various factors play into challenging waste management in the north, including minimal waste management services in northern Saskatchewan (Ghosh et al., 2023), low population density (Sebastian & Louis, 2022), and low household density, seasonal variation in house waste generation and low participation in household recycling (Lakhan, 2015). These factors contribute to significantly higher waste management costs, while contributing relatively little to overall diversion rates (Lakhan, 2015; Ghosh

DOI: 10.4324/9781003534129-8

& Ng, 2021). Lakhan (2015) questioned the sustainability of recycling programs in Ontario's north, especially considering economic and site-specific factors, noting that a balance in policy initiatives is likely to lead to more sustainable outcomes for an integrated waste management system in northern and remote areas. Collaborations among jurisdictions appears to benefit the efficiency of waste management systems (Ghosh et al., 2023; Richter et al., 2023). Ghosh et al. (2023) recommended that divisions that had at least one neighboring border with other provinces in Saskatchewan take a more active role in importing waste.

Sebastian and Louis (2022) proposed an integrated waste management system for all waste streams in the Northwest Territories, including difficult to manage wastes like tires, used oil, antifreeze, paint, and automobiles. In Saskatchewan, EPR programs exist for tires, used oil and antifreeze and paint, however, collection of these materials is sparse in the northern part of the province (SWRC, 2024).

Waste management in First Nations communities

There is also gap in understanding the state of waste management in Canadian First Nations communities. However, recent studies have started to close this gap (Assuah & Sinclair, 2021, Wang et al., 2023; Assuah, 2023). Indigenous Services Canada (ISC), a department of the federal government, is responsible for environmental protection, including solid waste management on First Nations (Assuah & Sinclair, 2021; Wang et al., 2023). Solid waste management regulation that falls under ISCs jurisdiction is outdated and fails to capture aspects of a modern integrated solid waste management system (Assuah & Sinclair, 2021). Assuah and Sinclair (2021) point out that these flaws have led to an environmental regulatory gap for these communities. Given that waste management falls under ISC's jurisdiction, Assuah and Sinclair (2021) found that confusion exists as to whether other waste management regulations (such as EPR regulations) are enforceable or applicable in First Nations communities. In fact, Assuah and Sinclair (2021) found that, in some circumstances, First Nations communities were told that they did not qualify to be part of EPR programs.

Assuah and Sinclair (2021) found that barriers to solid waste management on First Nations preclude them from being able to participate in EPR programs. These factors include preparedness or readiness of the community to participate in programs, availability of facilities to store waste prior to collection, willingness of EPR program operators to operate in remote areas, difficulty locating stewards for products, and general capacity of the community. Other factors such as economies of scale and the cost of transporting recyclables from remote locations are also of concern (Assuah & Sinclair, 2021; Haupt et al., 2018; Lakhan, 2015). This highlights the need for cooperation between stewardship agencies in northern, rural, and remote communities, to increase the presence of these programs. An important conclusion by Assuah and Sinclair (2021) is that EPR needs to be given a focus and attention in First Nations as it has been done in municipalities and provinces.

Bill S-5

Another important aspect of waste management in northern, remote, and First Nations communities is Bill S-5. Bill S-5 amends the Canadian Environmental Protection Act (CEPA), recognizing that all Canadians have a right to a healthy environment (Craik, 2023). It is important to note that this legislation doesn't bind provincial governments, but the amendment indicates strong movement toward environmental protection (Craik, 2023) for all Canadians. Environment and Climate Change Canada (ECCC) states that the bill will require decisions made under CEPA to respect the right to a healthy environment (ECCC, 2023). Furthermore, the bill explicitly accounts for principles of environmental justice, nonregression and intergenerational equity (ECCC, 2023). Environmental justice, the avoidance of adverse effects that disproportionately affect vulnerable populations (ECCC, 2023), is particularly interesting in the case of waste management on First Nations. A lack of proper waste management and recycling infrastructure on First Nations land may lead to higher pollution in these communities. Using EPR approaches offers an efficient solution to improve recycling on First Nations, and would reduce the likelihood of improper waste disposal which leads to pollution.

Saskatchewan EPR regulations

In Saskatchewan, *The Environmental Management and Protection Act* and *The Municipal Refuse Management Regulation* govern solid waste management. These pieces of legislation, or regulations enacted under them, do not apply to First Nations, since First Nations fall under ISC's jurisdiction (Wang et al., 2023). Nonetheless, the government can still provide guidance on the operation and management of relevant facilities (Wang et al., 2023). In fact, Saskatchewan's Solid Waste Management Strategy addresses how to work together with First Nations communities to improve solid waste management (Wang et al., 2023).

All of Saskatchewan's EPR regulations have a requirement to provide accessibility in "all areas of Saskatchewan" (Government of Saskatchewan, 2019). This condition may be left purposely vague to be nonprescriptive. Making accessibility clauses too prescriptive may unintentionally increase costs for producer responsibility organizations (PROs), which may eventually be passed on to the consumer, and may undermine the effectiveness of recycling programs. PROs are business entities that collaborate with those identified in the regulations as responsible for the waste they put on the market, helping them to fulfill their regulatory obligations as set out in EPR regulations (Alberta Recycling, 2024). They are responsible for a variety of different aspects of the recycling program, one of which is providing collection services for materials that fall under their mandate.

In order to have an efficient and effective recycling systems, a balance needs to be struck between accessibility, cost, and emissions related to collection. According to Stewardship Agencies of British Columbia (SABC, 2021), accessibility to collection facilities or services must also consider size, weight, and the generation rate of the product, among other characteristics of the goods (i.e., durable versus nondurable), the hazards and economics of collecting, storing and transporting each product. Community characteristics such as size and proximity to a collection service, as well as the desire of the community to support sustainable solutions and health and safety are also important considerations (SABC, 2021).

The information presented thus far highlights some of the issues faced in northern, remote, rural and First Nations communities when it comes to waste management and recycling. The objective of this study was to carry out a mapping exercise to understand what reasonable coverage for EPR programs looks like in the Northern Saskatchewan Administration District (NSAD). The resulting map will show areas where coverage is required to reasonably meet the regulatory requirement of EPR regulations in the province. This map can be used by stewardship organizations to gain insight into strategically planning locations for collection depots and pick-up services.

Case study: Accessibility in the NSAD

Understanding waste management regionalization and developing data-driven methods to improve regionalization is increasingly popular in Canada (Karimi et al., 2022; Richter et al., 2019). Richter et al. (2019) proposed a novel method using recursive Thiessen polygons to optimize waste management regions in Saskatchewan. Results suggested that regionalization may not be the best approach in the NSAD due to low population density and limited road network, and rather, that different approaches may be required (Richter et al., 2019). Richter et al. (2021) proposed a method for regionalization in Nova Scotia using standard deviational ellipses, which considered collection and transportation in rural areas. The results suggested a need to understand the corelationship between the cost to recycle and the ability to provide recycling in rural areas to optimize regional waste management efficiency (Richter et al., 2021).

Karimi et al. (2022) and Richter et al. (2021) further built on this work by combining remote sensing data to improve the analysis and further optimize the development of waste management regions. Karimi et al. (2022) used nighttime light imagery to identify population centers in rural areas, helping to develop a data-driven waste management system on a regional scale. These studies specifically look at waste facilities (landfills and transfer stations), which helped to further contribute toward improving the entire integrated solid waste management system.

Study area

Figure 5.1 shows the province of Saskatchewan, outlined in gray (Sask Interactive Mapping, 2024). Highways (GeoHub, 2021) and First Nations Reserves (Natural Resources Canada, 2024) are also mapped. The dark blue border indicates the boundary of the NSAD (Sask Interactive Mapping, 2024). All populated places in the NSAD are shown as blue dots (Statistics Canada, 2022a, 2022b). There are 52 populated places in the NSAD including First Nations communities, with a

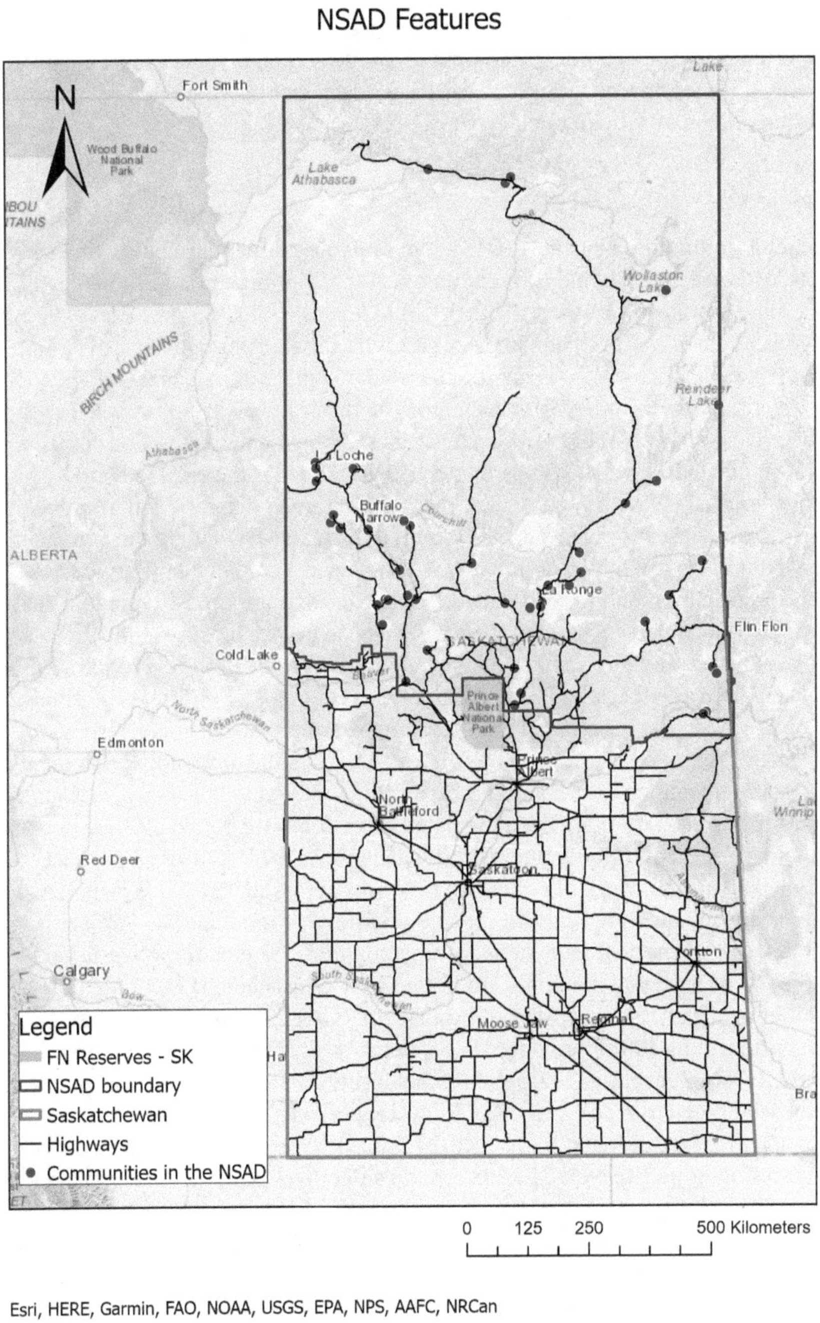

Figure 5.1 Study area showing Saskatchewan, North Saskatchewan Administration District and populated places within the NSAD, First Nations reserves, and road network.

Sources: Provincial boundary (Sask Interactive Mapping, 2024); Highways (GeoHub, 2021); First Nations Reserves (Natural Resources Canada, 2024); NSAD boundary (Sask Interactive Mapping, 2024); populated places (Statistics Canada, 2022a, 2022b).

population of 35,986, only 3 percent of the provincial population (Government of Saskatchewan, 2024). Given the population density and sparse road network, there lies an opportunity to make data-driven decisions to strategically improve collection efficiency (Richter et al., 2019, 2021, 2023).

Method

Geographic information systems (GIS) continue to be a popular choice for many aspects of waste management, including regional optimization (Richter et al., 2019), route optimization (Vu et al., 2019), and landfill siting (Karimi et al., 2020). In this study, the "Create Drive-Time Areas" tool, available through ArcGIS Online, is used to develop drive-time areas for all populated places in the NSAD. The tool uses a street network to determine the area of reach for a certain travel time or distance (Esri, 2024). Unlike the Create Buffers tool, which creates a buffer using straight-line distance, this tool has the advantage of using the road network. This is particularly advantageous given the nature of the road network and population spread in northern Saskatchewan. The required inputs of the tool are the populated places (point layer) and the required travel distance or time. The Esri Service Areas contains information on the road network and is used as an input for the tool. The output is a polygon layer showing the areas that can be reached given the inputs.

As mentioned previously, the regulated requirement of PROs is to provide access in "all areas of Saskatchewan." Given this vague requirement, different PROs have developed different accessibility measures. For example, Call2Recycle, responsible for the collection of consumer batteries, uses a 15 km radius from collection points to residential locations to measure accessibility (Call2Recycle Canada, 2019). In contrast, the used oil program has a regulated set of zones for both collection depots and alternative returns (Government of Saskatchewan, 2024). Differences in accessibility lead to a patchwork of collection networks across the province, which may be confusing for those wishing to recycle a variety of products. Most Canadian provinces and territories have stewardship programs for electronic waste, however the list of accepted materials are not entirely consistent (Hasan et al., 2024; Tasnim et al., 2024). It is worth noting that SARCAN, responsible for beverage container recycling in the province, has a network of 73 depots across the province (SARCAN Recycling, 2024). SARCAN locations also accept paint, batteries, and household electronics (SARCAN Recycling, 2024), and some locations may accept other recyclable materials as well. This depot system is similar to those used in British Columbia (RecycleBC, 2024) and can help to improve the convenience of recycling.

Stewardship Agencies of British Columbia (SABC) created an accessibility framework with the aim of providing guidance to determine how to provide a "reasonable" level of service across British Columbia (SABC, 2021). The SABC framework was specifically developed in consideration of British Columbia's challenging size and population distribution (SABC, 2021), both of which are similar in northern Saskatchewan. SABC (2021) defines the following criteria for accessibility:

- Service provided within 60 km for nonurban communities
- Service provided within 15 km for urban communities, and
- Where practical for remote communities.

Where "urban" is considered all areas within a Census Metropolitan Area (CMA) as defined by Statistics Canada (SABC, 2021). "Nonurban" areas are all those that are not CMAs. Further, a remote community is a nonurban community that is situated far from main centers of population without reliable infrastructure or those that are accessed using alternate mechanisms (SABC, 2021) such as fly-in/fly-out.

In this study, a 60 km radius was adopted, in line with the accessibility criteria for nonurban communities outlined by SABC (2021). The radius was applied to all populated places in the NSAD. The authors acknowledge that this radius may not be practicable in every northern remote community, or communities that are fly-in/fly-out.

Results

Figure 5.2 shows the result of running the "Create Drive-Time Areas" tool for a 60 km distance from each populated place in the NSAD. The results suggest there are a very specific set of areas in the NSAD that require collection. Five polygons (shown in purple) were created through the analysis, while the remaining areas lie solely along existing road network. These polygons likely indicate areas where there is a higher road density, which may correlate with a more concentrated pocket of population. The five identified areas (purple polygons) are likely to be good candidates for collection locations due to their proximity to population centers and a relatively denser road network. PROs can use this map to strategically place collection depots to capture materials as efficiently as possible. Where possible, a collaborative approach between PROs is recommended, since it is likely to alleviate issues with capacity in these communities, a barrier noted by Assuah and Sinclair (2021).

Readers will notice that some populated places on the map do not have any resulting drive-time area, for example, the three most northern populated places in the province (Fond-du-Lac, Stony Rapids and Chicken IR 224). These locations may be fly-in/fly-out or may lack a larger formal road network in the area. Waste management and recycling in these areas is likely to require a specific approach, since other means of waste management may not be feasible.

The results further suggest that interprovincial cooperation may be required, since some communities appear to be more accessible through Manitoba. See, for example, the large drive-time areas in Flin Flon and Sturgeon Landing areas on the Saskatchewan/Manitoba border, shown in Figure 5.2. Richter et al. (2023) found that interprovincial cooperation for waste management between Saskatchewan and Manitoba may be advantageous, especially near the border. Similar findings were reported by Ghosh et al. (2023). The results from this study suggest that the same is true for recycling. However, given that environmental handling fees (EHFs) may

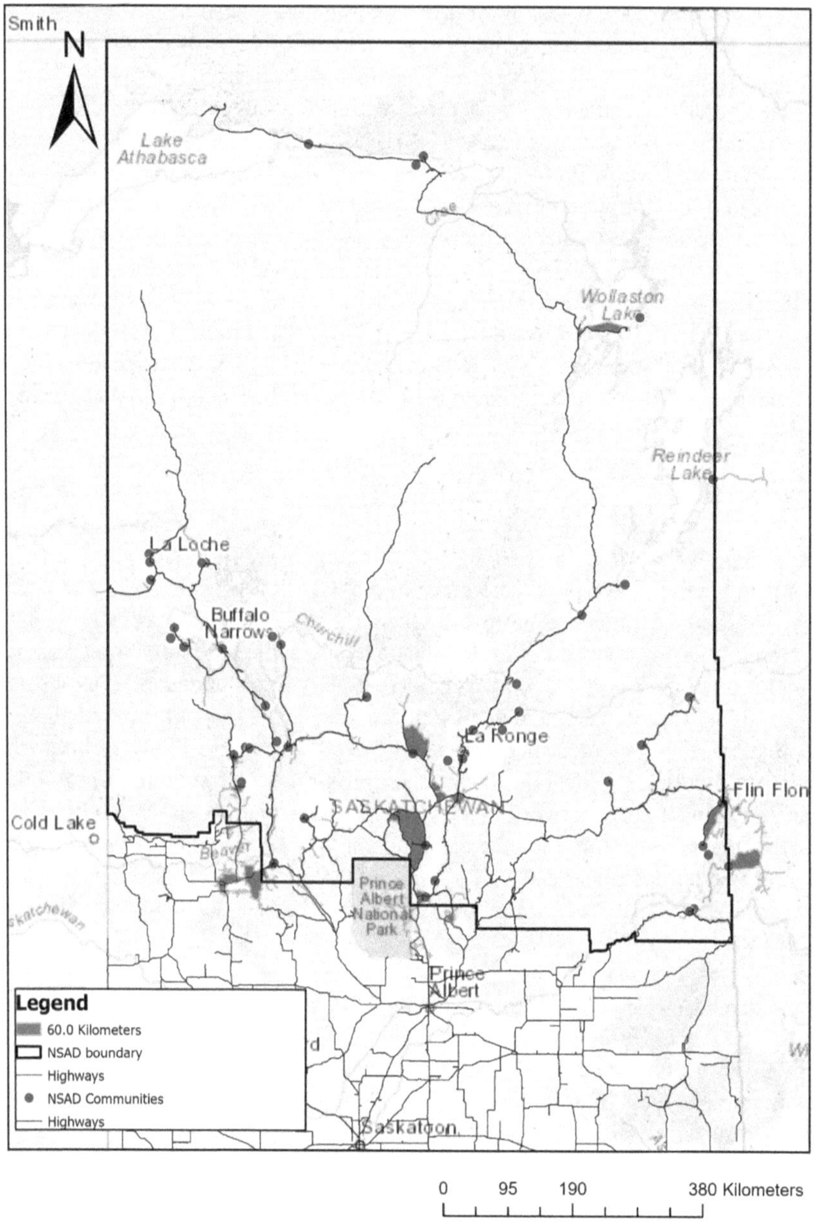

Figure 5.2 Drive-time areas (60 km) from all populated places in the NSAD.

vary by province, and are charged at the point of sale, the administration required to manage interprovincial cooperation for regulated recycling programs may be complex. This further highlights an advantage of consistent approaches to EPR across Canadian jurisdictions, an initiative highlighted by the Canadian Council of Ministers of the Environment (CCME, 2022).

While there are some clearly advantageous locations for PROs to locate or concentrate their collection efforts, there are a number of other locations that may still lie out of reach. On the map (Figure 5.2), these tend to be locations that don't have polygons, but rather, the analysis only highlighted a stretch of road. These communities will require a more specialized approach for collection (Richter et al., 2019). Given that these areas likely have sparse population, a more thorough investigation of waste generation in these areas is warranted to understand recycling demand. CleanFarms, responsible for the collection of grain bags in Saskatchewan, is currently testing out a pilot initiative to collect unwanted pesticides and livestock/animal medications (wastes which are not currently included in regulation; CleanFarms, 2024). They alternate collection back and forth between northern and southern Saskatchewan every second year (CleanFarms, 2024). Collecting waste every other year is efficient from a program perspective, while also ensuring timely collection of these materials. A similar approach could be used for all stewarded wastes in the north. However, it cannot be overstated that a thorough understanding of waste generation and composition is required to inform the development of such a program.

Considerations, limitations and future work

The result of this study is meant to provide insight into the design of a collection system in northern, remote, and rural areas. However, there are many other considerations and inputs for an optimized integrated waste management system. The authors acknowledge that looking at driven distance is only one part of a complex system. Factors such as road conditions and closures due to weather conditions, topographical features, and other considerations that may specifically impact the ability to travel in the NSAD have not been considered in this study. A full understanding of limitations that communities face is required. Furthermore, addressing issues related to data availability in First Nations is key. There is very little information available on solid waste generation, management, and disposal in these communities (Assuah & Sinclair, 2021; Wang et al., 2023). Furthermore, this GIS analysis does not consider socioeconomic factors such as differences in living standards or climactic conditions.

Remote communities don't exist solely in the North Saskatchewan Administration District. Southern Saskatchewan has many remote communities as well. Understanding proximity is key to developing an integrated system for waste management across the province. This idea can be further extended to Canadian First Nations. Wang et al. (2023) found that 5% of First Nations people in Canada are located in a zone more than 350 km from the nearest service center. As such, an investigation into accessibility for First Nations across Canada could result in interesting findings that could help to improve access to recycling and other waste management initiatives for these communities.

Conclusion

This study analyzed the challenges and opportunities associated with waste management in northern, remote, rural and First Nations communities in the North Saskatchewan Administration District. The result of the case study was a map showing locations where it may be more advantageous to place depots to collect recycling in the NSAD. The policy implications from the resulting map highlight the importance of considering the environmental burdens associated with collection and recycling, and further highlight that understanding optimal recycling rate is key, similar to findings from previous studies (e.g., Lakhan, 2015; Haupt et al., 2018). Policy approaches that support cooperation and collaboration among PROs or the development of recycling depots that collect multiple different waste streams may be advantageous.

Remote, rural, and First Nations communities face a variety of different challenges when it comes to waste management. Interplay between accessibility, regulatory requirements, and social and environmental aspects add further complexity. Regulated recycling programs in the province are required to provide access to recycling in "all areas of Saskatchewan," however, given the sparse population it can be difficult to do so economically. Using GIS, the "Create Drive-Time Areas" tool was leveraged to create a map that PROs can use to strategically place collection depots. The results showed that there were five locations that would likely make good candidates for collection depots. However, a different approach may be required for areas that are not covered by these five polygons. A unique approach, such as that employed by CleanFarms, where collection alternates between northern and southern parts of the province. could be used but would require a better understanding of waste generation in those areas. The results also suggest that interprovincial cooperation may be useful, however, administrative issues may make this very difficult to carry out. Ultimately, the mapping exercise provides a first step in the design of a collection system in for northern, remote, rural and First Nations communities. The authors acknowledge that the system is complex, and without a better understanding of waste generation and composition in these areas, it is difficult to design a proper integrated waste management system.

References

Alberta Recycling. (2024). *The role of producer responsibility organizations (PROS) in EPR*. www.albertarecycling.ca/the-role-of-producer-responsibility-organizations-pros-in-epr/

Assuah, A. (2023). "A lot of people ignore our culture when it comes to waste management": Examining the impacts of culture on solid waste management in two Canadian First Nations. *AlterNative: An International Journal of Indigenous Peoples*, 19(2). https://doi.org/10.1177/11771801231163635.

Assuah, A., & Sinclair, A. J. (2021). Solid waste management in western Canadian First Nations. *Waste Management*, 129, 54–61. https://doi.org/10.1016/j.wasman.2021.05.007.

Call2Recycle Canada. (2019). *Saskatchewan product stewardship plan for consumer batteries*. Accessed from www.call2recycle.ca/saskatchewan/ on June 27, 2024.

Canadian Council of Ministers of the Environment. (2022). *Guidance to facilitate consistent extended producer responsibility policies and programs for Plastics*. Accessed from https://ccme.ca/en/res/eprguidanceen.pdf on June 27, 2024.

CleanFarms. (2024). *Saskatchewan programs & events*. Accessed from https://cleanfarms.ca/programs-at-a-glance/sk-programs-events/ on June 27, 2024.

Craik, N. (2023). *Q and A with the experts: Canada's new environmental legislation*. University of Waterloo Media Relations. Accessed from https://uwaterloo.ca/news/media/q-and-experts-canadas-new-environmental-legislation on June 25, 2024.

Environment and Climate Change Canada. (2023). *Bill S-5: Strengthening environmental protection for a Healthier Canada Act: Backgrounder*. Accessed from www.canada.ca/en/environment-climate-change/news/2023/06/bill-s-5-strengthening-environmental-protection-for-a-healthier-canada-act.html on June 25, 2024.

Esri. (2024). *Create drive-time areas*. ArcGIS Online Documentation. Accessed from https://doc.arcgis.com/en/arcgis-online/analyze/create-drive-time-areas.htm on June 25, 2024.

GeoHub. (2021). Saskatchewan highway network shapefile. Accessed from https://geohub.saskatchewan.ca/datasets/ on June 27, 2024.

Ghosh, A., & Ng, K. T. W. (2021). Temporal and spatial distributions of waste facilities and solid waste management strategies in rural and urban Saskatchewan, Canada. *Sustainability*, 13(12), 6887. https://doi.org/10.3390/su13126887.

Ghosh, A., Ng, K. T. W., & Karimi, N. (2023). An evaluation of the temporal and spatial evolution of waste facilities using a simplified spatial distance analytical framework. *Environmental Development*, 45, 100820. https://doi.org/10.1016/j.envdev.2023.100820.

Government of Saskatchewan. (2018). The used petroleum and antifreeze products stewardship regulations. Accessed from https://usedoilrecyclingsk.com/wordpress/wp-content/uploads/2019/04/E10-22R7.pdf on June 27, 2024.

Government of Saskatchewan. (2020). *Saskatchewan's solid waste management strategy*. Accessed from www.saskatchewan.ca/residents/environment-public-health-and-safety/saskatchewan-waste-management/solid-waste-management-strategy on August 6, 2024.

Government of Saskatchewan. (2024). *Northern Saskatchewan Administration District*. Accessed from www.saskatchewan.ca/business/first-nations-metis-and-northern-community-businesses/economic-development/northern-administration-district on June 26, 2024.

Government of Saskatchewan, Ministry of Environment. (2019). *The household hazardous waste products stewardship regulations (Chapter E-10.22 Reg 8)*. Accessed from https://publications.saskatchewan.ca/#/products/101719 on June 26, 2024.

Hasan, M. M., Mahmud, T. S., Assuah, A., Ng, K. T. W., Tasnim, A., & Abha, A. T. (2024). An investigation on the operational resilience of the Canadian electronic product stewardship program and the recycling business characteristics. *Waste Management*, 181, 68–78. https://doi.org/10.1016/j.wasman.2024.04.002.

Haupt, M., Waser, E., Würmli, J. C., & Hellweg, S. (2018). Is there an environmentally optimal separate collection rate? *Waste Management*, *77*, 220–224. https://doi.org/10.1016/j.wasman.2018.03.050.

Karimi, N., Ng, K. T. W., & Richter, A. (2022). Integrating geographic information system network analysis and nighttime light satellite imagery to optimize landfill regionalization on a regional level. *Environmental Science and Pollution Research*, 29, 81492–81504. https://doi.org/10.1007/s11356-022-21462-w

Karimi, N., Richter, A., & Ng, K. T. W. (2020). Siting and ranking municipal landfill sites in regional scale using nighttime satellite imagery. *Journal of Environmental Management*, 256, 109942, 1–9. https://doi.org/10.1016/j.wasman.2019.11.028.

Lakhan, C. (2015). North of the 46° parallel: Obstacles and challenges to recycling in Ontario's rural and northern communities. *Waste Management*, 44, 216–226. https://doi.org/10.1016/j.wasman.2015.06.044.

Natural Resources Canada. (2024). *Indian reserve feature layer*. Accessed from https://gis.saskatchewan.ca/arcgis/rest/services/AboriginalLands/MapServer on June 27, 2024.

RecycleBC. (2024). *Find a depot*. Accessed from https://recyclebc.ca/where-can-i-recycle/find-a-depot/ on June 27, 2024.

Richter, A., Ng, K. T. W., & Karimi, N. (2021). Developing a novel proximity analysis approach for assessment of waste management cost efficiency in low population density areas. *Sustainable Cities and Society*, 65, 102583, 1–12. https://doi.org/10.1016/j.scs.2020.102583.

Richter, A., Ng, K. T. W., & Karimi, N. (2023). Optimizing waste management regions spanning inter-provincial boundaries. Proceedings of the Canadian Society of Civil Engineering Annual Conference 2021, *Lecture Notes in Civil Engineering*, 249, 121–132. https://doi.org/10.1007/978-981-19-1061-6_13.

Richter, A., Ng, K. T. W., Karimi, N., Wu, P., & Kashani, A. H. (2019). Optimization of waste management regions using recursive Thiessen Polygons. *Journal of Cleaner Production*, 234, 85–96. https://doi.org/10.1016/j.jclepro.2019.06.178.

SARCAN Recycling. (2024). *How to SARCAN*. Accessed from https://sarcan.ca/ on June 27, 2024.

Sask Interactive Mapping. (2024). *Northern Saskatchewan Administration District*. Accessed from https://gisappl.saskatchewan.ca/Html5Ext/index.html?viewer=saskinteractive on June 27, 2024.

Saskatchewan Waste Reduction Council. (2024). *Waste reduction hub*. Accessed from www.saskwastereduction.ca/ on June 27, 2024.

Sebastian, R. M., & Louis, J. (2022). Waste management in Northwest Territories, Canada: Current practices, opportunities, and challenges. *Journal of Environmental Chemical Engineering*, 10, 106930. https://doi.org/10.1016/j.jece.2021.106930.

Statistics Canada. (2022a). *Population and dwelling counts: Canada, provinces and territories, and census subdivisions*. Table 98-10-0002-02. https://doi.org/10.25318/9810000201-eng

Statistics Canada. (2022b). *Population and dwelling counts: Canada and populations centers*. Table 98-10-0011-01. https://doi.org/10.25318/9810001101-eng

Stewardship Agencies of British Columbia. (2021). *Guidance on accessibility of stewardship programs (accessibility framework)*. Accessed from www.stewardshipagenciesbc.ca/resources/ on June 11, 2024.

Tasnim, A., Chowdhury, R., Mim, S. J., Ng, K. T. W., & Adu-Darko, H. (2024). Influence of Canadian provincial stewardship model attributes on the cost effectiveness of e-waste management. *Journal of Environmental Engineering*, 358, 120945. https://doi.org/10.1016/j.jenvman.2024.120945.

Vu, H. L., Bolingbroke, D., Ng, K. T. W., & Fallah, B. (2019). Assessment of waste characteristics and their impact on GIS vehicle collection route optimization using ANN waste forecasts. *Waste Management*, 88, 118–130. https://doi.org/10.1016/j.wasman.2019.03.037.

Wang, Z., Chen, Z., & An, C. (2023). A review on solid waste management in Canadian First Nations communities: Policy, practices and challenges. *Cleaner Waste Systems*, 4, 100074. https://doi.org/10.1016/j.clwas.2022.100074.

6 Municipal solid waste management in Northern Manitoba municipalities

Anderson Assuah

Introduction

Municipal solid waste (MSW) management continues to engender discussions and conversations locally, nationally, and globally, as the world grapples with increased waste generation and poor management practices. Approximately 2.24 billion tonnes of MSW were produced globally in 2020, out of which 13% was recovered and 87% disposed of (UNEP, 2023). With current generation rates and management such as landfilling and burning, this figure is estimated to reach about 3.8 billion tonnes in 2050 (UNEP, 2024). This global trend is similar in Canada, where MSW generation stood at 36 million tonnes in 2022, representing an increase of about 17% from 2000 (ECCC, 2024). While the amount of waste diverted increased over this period, landfilling remains the predominant method of managing MSW in Canada. Landfills contributed to 23.9% of Canada's methane emissions in 2020 (Statistics Canada, 2022a).

Three levels of government share responsibility for managing MSW in Canada. The federal government oversees international and interprovincial/interterritorial movement of hazardous waste and establishes best practices to reduce greenhouse gases and pollution; provincial and territorial governments develop plans and programs, approve and monitor facilities and operations, and ensure compliance or enforce applicable laws; and, municipal governments are responsible for collection and management, such as disposal, incineration, recycling, and composting (ECCC, 2024). As Canadian provinces develop plans and strategies to curb increasing MSW generation and adopt more appropriate management methods, it is important that rural, northern, and remote regions are given the needed attention and resources.

This is because the literature suggests that MSW management services are generally poor in these regions when compared with urban areas (Sebastian & Louis, 2022). Lack of infrastructure and facilities, inadequate funding, low population densities, and unavailable MSW services and programs are some of the key challenges these regions face (Lakhan, 2015; Sebastian & Louis, 2022; Assuah, 2023). Unique characteristics of northern communities, such as weather, remoteness, and sparse populations also contribute to the challenges that communities face. Additionally, Ghosh and Ng (2021) argue that research is skewed toward investigating waste management issues in urban centers as against rural areas. These

DOI: 10.4324/9781003534129-9

create a gap in the literature regarding understanding MSW management habits, characteristics, and management in northern communities. Among the resulting impacts of this are lack of data for sound policy development and decision-making, lack of baseline data for researchers to interrogate pertinent MSW issues, and lack of information for learning among municipalities.

There is a gap in the literature and understanding of MSW management in Manitoba, particularly in northern Manitoba—the region generally described as north of the 53rd parallel. For example, it unknown how much MSW northern Manitoba communities contribute to the approximately 900,000 tonnes generated in 2022 (Statistics Canada, 2024) nor the amount of waste diverted. Again, there is no publicly available data and/or information on the MSW management programs being pursued by these communities, and the region's MSW characterization. In fact, the lack of data was evident in a national waste management study that depended on two studies from southern Manitoba municipalities—City of Winnipeg and City of Swan River—to examine waste characterization in the province (ECCC, 2020). Given the foregoing, the objective of this paper was to provide an overview of MSW management in northern Manitoba municipalities and offer insights into factors that need to be given attention.

Methods

This was a qualitative multiple case study research that gathered in-depth information from nine municipalities in northern Manitoba (Yin, 2018) about MSW management. Table 6.1 shows the municipalities and their population. The first two municipalities are classified as cities and the remainder are towns (Association of Manitoba Municipalities [AMM], 2024). Mystery Lake is a local government district, which is regarded as a municipality (AMM, 2024); however, the population of the area is zero according to Statistics Canada (2023). As a result, this area was excluded from the research.

Semi-structured interview was the data collection method utilized to gather information from nine MSW managers, recycling managers or any person with

Table 6.1 Northern Manitoba municipalities and population

Community	*Population*
Thompson	13,035
Flin Flon	4,940
The Pas	5,639
Snow Lake	1,088
Gillam	1,007
Churchill	870
Lynn Lake	579
Leaf Rapids	351
Grand Rapids	213

Source: The information in the table was sourced from Statistics Canada (2023).

similar portfolio, and a Chief Administrative Officer. These participants were purposely selected because of their involvement in day-to-day operations and management of waste facilities and programs. Interview questions focused on topics, such as MSW management programs, effectiveness of programs, capacity building, and opportunities for recycling. All interviews were audio-recorded with permission from participants. Participants were given their interview transcripts to check them before the information provided were used.

The transcribed data were organized and coded using NVivo 14, and themes from the literature and those grounded in the data were developed. These themes are discussed in the next section of the paper, utilizing quotes from participants. The quotes used are those that typically represent the views of participants, except indicated. To comply with anonymity and confidentiality for the research, pseudonyms—P1, P2, etc.—were used to represent participants, and the names of municipalities were not associated with the information provided. This research was approved by the University College of the North Research Ethics Board (UCN 2023/24-INT-02).

Results and discussion

Solid waste management infrastructure and programs

Participants indicated that engineered landfills, dumpsites, and recycling depots/waste transfer stations are the infrastructure in their municipalities (Table 6.2). All municipalities had an engineered landfill or dumpsite; however, only six had multi-material recycling infrastructure. The existence of more disposal facilities than waste diversion infrastructure is consistent with the findings that more of the former was built than the latter in rural Canadian municipalities between 2010 and 2020 (Statistics Canada, 2022). This was the opposite for urban municipalities in this same period.

Similarly, all municipalities had community-wide household garbage collection, but only six run a community-wide household recycling program (Table 6.3). None of the municipalities had a community-wide composting nor reuse program. Curbside pick-up and drop-off by residents were the two common ways garbage is collected in municipalities except for two of them, who only provide curbside pick-up. Municipalities with household recycling

Table 6.2 Types of MSW infrastructure in municipalities

Types of infrastructure	*Communities*
Engineered landfill	5
Dumpsite	4
Recycling depot/waste transfer station	6
Composting	0

Table 6.3 Community-wide MSW management programs

Programs	*Number of communities*
Household garbage disposal	9
Household recycling	6
Composting	0

programs also collect waste through these avenues, and collection ranges from one to five times per week. Approximately 79% of the 65% of Canadians, who composted kitchen waste in 2021, did so through curbside pick-up programs (Statistics Canada, 2024). However, the infrastructure does not exist in northern Manitoba municipalities for composting. In fact, the province has one of the lowest composting rates in Canada, as one in three households composted their kitchen waste in 2019 (Statistics Canada, 2022b). There is great need to increase MSW diversion infrastructure (e.g., composting infrastructure) in northern Manitoba to encourage and increase diversion programs.

Five of the six municipalities that provide recycling services assist surrounding communities. "The Recycling Centre is the only game in town … There are also quite a few people who live out at the lakes, and they use our recycling. They have a separate landfill, but they use the Recycling Centre" (P3). P4 also indicated, "Yes, we serve other areas for a fee." For their part, P8 explained that their municipality "operates a materials recovery facility and therefore process recyclables from Thompson, First Nation communities, provincial parks such as Paint Lake, outlying communities, and hydro generating stations." Cans, glass bottles, glass jars, plastic bottles, paper, and cardboard are examples of recyclables that are collected.

Some participants mentioned that they collect their comingled waste and send them to other northern facilities to sort them out. "We have co-mingled recycling, so the cardboard, bottles, and cans all go together into the shipping bags, and then we send them to Gardewine. Gardewine takes it to the Thompson recycling depot, they sort it, and we get our rebate," said P1. P5 added, "Our recycling is gathered up and bagged in big bags … We do not actually handle it. We do not sort it out, we ship it to Thompson." These findings suggest that some of the municipalities are working with those with better infrastructure to successfully deliver their programs. Given the many challenges that northern communities face with MSW management, such as sparse populations and harsh climatic conditions and weather (Arctic Council Secretariat, 2024; ECCC, 2017), finding ways to work together could help improve management of programs in the region.

The three participants whose municipalities do not have recycling programs mentioned lack of support from their municipalities for recycling, lack of interest and participation from residents, high operational cost, and lack of infrastructure as reasons for nonexistence of programs. A participant explained:

> We used to do recycling about maybe 10 years ago, but we got rid of it. We decided we were going to go just straight to landfilling everything … When we were recycling, I think people did it because they had to do it. I do not think it was as much a voluntary thing … I think people are realizing there is only so much you can do, and it is costing a lot of money [to recycle].

The above factors highlight the need for careful planning when communities decide to pursue programs. Components of such plans should include waste audits, establishing priorities, developing a plan, and building feedback loops to evaluate and improve programs (ECCC, 2017; Federation of Canadian Municipalities, 2017). These processes should actively involve community members, so they have "ownership" of the programs to increase their participation.

It is noteworthy that all municipalities sort out waste, such as electronics and white goods (e.g., fridge and microwave) at their landfill or dumpsite, including those without household recycling programs.

Building capacity

Community education

Five participants explained that they have and are actively utilizing a variety of means to educate and get residents involved in their programs, while the remainder did not have any such approaches. For the latter group, very basic information such as opening and closing times are provided. "No, nothing like that [building capacity]. The only thing we will do is inform residents when we are open and closed, but other than that, no, we do not" (P6).

For the former, building capacity included making Facebook posts, distributing flyers, radio advertisements, and open houses to discuss programs. While some participants mentioned that they do these regularly, it was not often for others. P2 provided the following details:

> We regularly put out public notices. We have had a couple of open houses here in town for community engagement, and like I said, we are going to be doing another one next week to inform the community and get them on board. We are going around and talking to the major producers of waste, like the restaurants, about composting and separating some more of these materials out of the composter.

P8 added:

> We participate in a trade show every year, and we have flyers with information on the recycling center. We also have a Facebook account, and prior to the Covid-19 pandemic, we were doing presentations in the schools; we are going to start that up again. We also have a video of our recycling center. I believe

> they had an open house here, but with Covid-19, all of that stopped. I think we are going to get that up again in the summertime because it would be nice for people to come and see what is done here and that we sort out everything by hand. Nothing is done by machine here.

Building capacity in the many ways described above gives residents the opportunity to participate in decision-making and provide feedback where necessary; however, it is important to gauge the effectiveness of these methods. Participants had mixed reactions about this, as some of the approaches have proven more effective than others. As P2 indicated:

> The only ones we really get feedback from is the community engagements, where we get them all into a room and people come by and discuss. Like I said, this last one had a bit less people. It was not a complete failure, but it was disappointing to only have those 26 people show up.

P3 explained how they use Facebook to encourage residents to read about their newspaper article:

> I sometimes wonder how many people read my newspaper column. I mean, just because a lot of people do not read it, I used to put them up on the website, but I am about a year behind that. I know there are people in town that read it because they tell me, or they will comment on something that was in there. I use Facebook to say check the paper tomorrow because I am discussing this or here is a picture of this or here is something else I wanted to say … So sometimes, you need to put a little extra, but I hope that gets people looking at the newspaper.

It is important that there are various avenues that residents can learn and build their capacities, because MSW management requires careful and continuous learning by all stakeholders to adjust to changes. Capacity building avenues that allow different stakeholders to make input into policies, plans, and decisions can be helpful to their participation (Gutberlet, 2016). While the outcome of these approaches is important, the processes involved in providing information, discussing issues, and training residents is equally important to enable learners gain much knowledge (Herat, 2015).

Staff training

Six participants mentioned that their staff have been trained in operating forklifts, Workplace Hazardous Materials Information System, and safety training, while there has not been any training for three participants and their staff. P2 revealed that, "We have all gone through extensive training. Since I have been here, we have done training for recycling, MSW, and cardboard. We have all gone through quite a bit of training … I have gone to Winnipeg for training on MSW." P8 highlighted both formal and on-the-job training that they have provided to their staff:

> We currently have 8 permanent staff, and it is on hands training for the staff we have. We do provide staff training on the skid steer and health and safety courses, but majority of the training is hands-on. They get certified once, and then there are courses that they can take 2 years later that they can take to recertify … There are no other facilities around here, so we are completely dependent on training people regarding how to sort and the safety practices and line handling, sharp edges and broken glass.

On the contrary, P7 explained why there has not been training for their staff:

> No training. It would be like asking them if they wanted to take a course on washing your car. Waste management is a dump. How do we get our garbage out of the community, and how do we manage it well? Well, you take it out to the dump, and there are specific spots to put it. What do you need a course for? That would be the mindset.

This latter participant assumes that workers automatically should know how to handle and manage waste properly and adhere to all environmental health and safety that comes with the job, but this is not the case. The Solid Waste Association of North America (2024), for example, reports that in Canada and United States, there was an increase in fatalities to solid waste workers in 2022 when compared to 2021. This report also revealed that mechanical-related fatalities (e.g., truck maintenance and work on shredders and balers) and workers struck by waste vehicles as two leading causes of fatalities. Therefore, training for waste collectors/managers should be a priority to communities.

Funding

MSW management requires significant financial investments to operate facilities and run programs. Participants provided information on their sources of funding for operations. All participants indicated that, generally, they receive funding from their municipalities. Grants, fee structure for garbage disposal, and rebates from producer responsibility organizations (PROs) were other sources mentioned. According to P7, operations are funded through

> tax revenues and fee structure … We receive different municipal grants, which are then used at the discretion of the municipality. Let's say, it is $40,000. We use it as far as our operating expenses. Within our operating expenses is the waste management site, so we do receive provincial funds that generally can be used across the budget as we see fit.

Participants who utilize rebates to support their programs explained that, "When we sort our recycling, we get money for our aluminum and plastics, and all depends on the market for them. Then we take lead batteries, and we get money for that. Money from household hazardous waste and electronics" (P4).

P8 explained the various sources of revenue they receive for their programs and their impact:

> The city gives us a grant each year. We get revenue from Multi-Material Stewardship Manitoba (MMSM), and we have some revenue coming from our commercial pickups. I would say they probably pay for themselves; they are not a real revenue generator … Cans generate revenue, cardboard used to do but not now. Most of the materials are negative [costs] when we have to pay to have them processed and or when we have to ship them out … They charge us a fee per tonnes for processing our commingled materials … Aluminum cans are not drawing in the value they used to, and the cardboard, if we are lucky, balances it, but basically, by the time we sort the cardboard and package it, we are lucky to get enough to make the shipping cost. So, the product revenue stream, is minor. In fact, it is negative, and we are dependent on the city and Multi-Material Stewardship Manitoba.

Provincial stewardship programs

In Manitoba, there are 12 PROs that deal with the end-of-life of waste (Manitoba Conservation and Climate, 2021). The data on the number of PROs municipalities work with was mixed, as there were municipalities that deal with most of the PROs, with few, and two that work with none. For municipalities that manage recycling and landfill differently, management under each is responsible for working with PROs. For instance, P8 revealed that their recycling facility works with MMSM, Product Care, and Call to Recycle, while "scrap metal, e-waste, tires, and hazardous waste are the responsibility of the city [landfill]. They keep separate funding and management practices." Although they work with most of the PROs, P2 noted, "We have not done anything for quite some time. I would say 4 or 5 years now … Product care and EPRA are our major ones right now."

Two participants mentioned that they have only worked with MMSM. This results from "a lack of awareness, in all honesty, because I did not know anything about it [PROs], and I am not sure if our office staff know anything else for the other programs" (P1).

MMSM started operations in 2010 (MMSM, n.d.) and this could explains the reason these two municipalities know about their operations and have been working with them. Contrarily, that they are not aware of any other PRO and two other municipalities do not work with PROs suggest a challenge with the operations of the PROs in northern Manitoba. This could be because the PROs have not actively worked to include all northern Manitoba municipalities in their operations and or municipalities have not made efforts to reach out to the PROs. This gap needs to be bridged because the mandate of PROs covers the entire province.

Municipalities that do not work with PROs or work with few of them, however, look for recyclers to purchase their materials. According to P6:

> We collect tires and then a guy comes up and collects them; I think they are from Thompson or Winnipeg. They come and collect them once or twice per year, depending on how much we have. For the metal, what we usually do is to hire a contractor, and we give them whatever money, so they can get our steel out rather than we pay them. That is how they get paid. So, we get a small company to come up, and they load all the material up, and they take all the materials, and whatever money they can get for that.

Generally, participants appreciate that they can work with the PROs to get their waste materials out from their communities. P3 explained this and revealed that MMSM is making changes to its program structure and payment.

> MMSM pays us. Theirs is a really challenging setup as they pay you based on your tonnage 2 years ago and on the value of materials, so you are kind of behind on what is happening with their payments, but they are still a very big part of what we do. And they are trying to move to what they call a full extended producer responsibility, so they will oversee all the costs upfront; nobody has to pay, and they will be running the programs instead of somebody else running them, and they pay them. But that is down the road for a little bit. But like, for instance, electronics, they pay the shipping, they pay us for handling it; same with batteries, they pay us for the shipping, they pay us for handling it, and that kind of stuff is big. I do not know how we could do without that.

MMSM currently reimburses municipalities up to 80% of the net cost of their recycling (MMSM, n.d.). Therefore, transitioning to a fully industry-funded model can increase participation in the program by municipalities and communities.

Challenges

Funding

Funding was the biggest challenge participants mentioned as impacting their operations, despite receiving financial support from various sources. P4 explained:

> Finding the funding to get new equipment is a challenge. Right now, we are using a truck and trailer to do curbside pickup, which is not ideal. To buy a garbage truck, we would need $300,000. Being non-profit, where do you get that money? Being non-profit, you cannot go to a bank to get a loan, so obtaining equipment is a real big challenge unless you can get government grants.

P6 also explained how lack of funding ended their recycling program.

> Funding is always a challenge. It is never there when you need it. You wish you could always plan for the unforeseen, but it is impossible … There is a challenge

> with funding when you want to get stuff from down south to the north and vice versa, everything is expensive. We reasonably closed our recycling program, because it came down to dollars and cents. So, you know, that is where the challenge is. If we ever wanted to have recycling going again, where do we get the money to do that?

MSW is a capital-intensive undertaking, for instance, building infrastructure, purchasing equipment, and yearly operational costs. These investments, by governments, do not usually yield profits. According to Statistics Canada (2021), "Local governments spent $3.5 billion on waste management in 2018 and reported revenues of $2.9 billion from it." This means municipal governments need a variety of revenue sources to adequately embark on their programs.

Staff availability and retention

All participants mentioned that a major challenge to their operations was hiring and retaining staff for their operations, particularly because of relatively low wages and the unavailability of people to work in MSW management. P2 explained:

> The only real challenge is, like I said before, not having staff that is able to get out to the facility to do the recycling. There have been some struggles for the most part … Given our small community and the workforce that we can draw from, it is difficult to find staff, and it is also difficult to get the funding to pay for these staff. So, we have typically, throughout the year, had 2 casuals who come in during the summertime start-up at the beginning of June and work through until the beginning of November. We also have a green team that comes in for July and August as well. But other than that, trying to find good quality staff in town is almost impossible.

The ability to pay fair wages to staff was highlighted by P8, who explained:

> One of the challenges we have is the limitation on the wages and benefits we can afford to pay our employees. It is considerably lower than those of municipal workers. Another challenge is staff turnover. There is abundance of jobs, which gives the employees opportunity to move around. Whereas before, it was the opposite, jobs were hard to come by and employees stayed.

Consequently, P4 declared, "You are not making a living wage out here anymore, or ever."

The small population size of municipalities (Table 6.1) coupled with low wages make them less competitive to attract interested and qualified staff. Municipalities would have to incentivize staff, for example, to retain them. Without such strategies, there will be reduced institutional knowledge in waste departments, and cost of operations would increase, since funds are needed to continually train new personnel.

Distance or remoteness

Some participants mentioned that the location of their municipalities means they are further away from major cities in Manitoba, which increases operational costs because of higher prices of supplies and makes it difficult to access supplies when needed. "The biggest thing is getting supplies and parts for our day-to-day operations—our vehicles, trailer, and the landfill. That would be the biggest challenge, just getting supplies and parts due to the distance" (P1). Additionally, distance from MSW management facilities increases transportation cost of programs and impact the sale of recycled materials. As P5 explained:

> Let's just take the scrap metal, for instance. I know the scrap metal is probably worth $80–$100 per pound, but the guy that is picking it up is paying me $40. Well, he is the only guy that is coming this far north with a vehicle to pick up. So, I pretty much must take the $40 that he is giving me and be happy with it. Because if I tell him no, then I have got nobody picking it up, so that is a challenge. Whereas down south, they have more scrap dealers, and you might get a better price for your recycled scrap metal.

P7 further explained how the economics does not favor doing MSW management business in the north:

> Our isolation is a challenge because we are isolated. And we are outside of the network, a profitable network. Like business models when they are developed, business analysis will look at distribution chain and infrastructure and see that okay, we could probably realise x amount of dollars per quarter based on this infrastructure. We do not have the infrastructure that any business analysts would be able to sit down and say, hey, we can profit at that. Unless we do our part that we create such a volume that fits their business model, we will not be making those dollars.

Distance away from major vibrant cities can increase cost of operations, delay getting items on time, and cut off communities from the recycling economy. Dealing with this challenge partly lies in communities pooling resources and working together on waste initiatives and programs that will benefit them. Some participants have already mentioned that they work with surrounding communities, and this can be replicated by other municipalities by formalizing working relationships to, for instance, reduce their cost. The other way of dealing with this situation is through policy at the provincial government level to deliberately create an economy for MSW management in northern Manitoba.

Attitudes and behaviors

All participants mentioned that despite their best efforts, some residents do not dispose of garbage appropriately at their facilities, while many others do not participate in the community-wide household recycling programs. For those that participate, there is still the need to inform and educate them to properly recycle. P3 extensively explained these challenges:

> The challenge is how to get the people who do not recycle to start. Some of it is that they just do not want to be bothered. A lot of them just do not think that it is important, or they are opposed to it. There are people who are opposed to it. And some of that is based on misinformation. The other big challenge is that you can put as much information out there all the time, but unless the person is in the right place at the right time and in the right frame of mind, they do not get it. You know, if they are not listening to the local radio at the right time, they are not going to hear my announcements. If they do not stop at the booth at the trade show, they do not get to see the pictures. If they just throw their flyers away, they will not get what is in the mailbox. If they do not care about recycling, they are not going to google recycling or look us up on Facebook. So, I think probably the big thing is trying to get people who do not actually feel like they should or could recycle to participate.

In addition, P5 mentioned: "People think that the status quo should be maintained. They do not see that a change would be good. They do not believe that they need to recycle, so why would they? So, that is a very big challenge."

Changing attitudes and behaviors about MSW management can be difficult because it takes a longer time and different interventions to achieve such results. This requires MSW managers to understand the system and programs they operate and concerns of residents. A one-size-fits-all approach would not be effective, as residents don't have equal understanding of programs.

Conclusion

This research has provided an overview of MSW management in municipalities in northern Manitoba (north of 53rd parallel), explaining factors including stewardship programs, waste infrastructure and programs, building capacity, funding, and challenges that impact daily operations and management. The data makes us to conclude that the nine municipalities examined are doing their best to manage programs (garbage and recycling) despite inadequate resources and challenges they face such as staff retention and isolation, which prevents them from fully benefiting from the business aspects of waste management.

Closing this gap requires developing a local economy and a northern Manitoba-focused strategy for MSW management. This strategy should involve municipalities collaborating and working together on initiatives and sharing resources. The involvement of the provincial government will be crucial in setting the tone and creating the enabling environment to achieve this. Additionally, PROs need to intensify their outreach to rope in municipalities and communities that have yet to be captured in their operations. This will require strengthening the capacities of municipalities and communities to meet program requirements.

Future research should investigate waste characterization in municipalities and the region, including assessing diversion rates. Additionally, examining the business models or aspects of MSW will be beneficial to understanding costs and benefits of operations and offer ideas into investments that municipalities need. Essentially, more research is needed on MSW management in northern Manitoba.

References

Arctic Council Secretariat. (2024, June 28). *Waste management in remote arctic communities: unique challenges and emerging solutions.* https://arctic-council.org/news/waste-management-in-remote-arctic-communities-unique-challenges-and-emerging-solutions/.

Association of Manitoba Municipalities. (2024, July 20). www.amm.mb.ca/members/municipal-map/northern.

Assuah, A. (2023). "A lot of people ignore our culture when it comes to waste management": Examining the impacts of culture on solid waste management in two Canadian First Nations. *AlterNative: An International Journal of Indigenous Peoples*, 19(2), 466–474. https://doi.org/10.1177/11771801231163635.

Environment and Climate Change Canada. (2017). *Solid waste management for northern and remote communities: Planning and technical guidance document.* www.canada.ca/en/environment-climate-change/services/managing-reducing-waste/municipal-solid/environment/northern-remote-communities.html.

Environment and Climate Change Canada. (2020). *National waste characterization report: The composition of Canadian residual municipal solid waste.* https://publications.gc.ca/collections/collection_2020/eccc/en14/En14-405-2020-eng.pdf.

Environment and Climate Change Canada. (2024, June 21). *Municipal solid waste management in Canada.* www.canada.ca/en/environment-climate-change/services/managing-reducing-waste/municipal-solid/environment.html.

Federation of Canadian Municipalities. (2017). Solid waste management toolkit. https://fcm.ca/sites/default/files/documents/resources/tool/solid-waste-management-toolkit-cipp.pdf.

Ghosh, A., & Ng, K. T. W. (2021). Temporal and spatial distributions of waste facilities and solid waste management strategies in rural and urban Saskatchewan, Canada. *Sustainability*, 13(12), 6887. https://doi.org/10.3390/su13126887.

Gutberlet, J. (2016). *Urban recycling cooperatives: Building resilient communities.* Taylor & Francis.

Herat, S. (2015). Waste management training and capacity building for local authorities in developing countries. *Waste Management & Research*, 33(1), 1–2. https://doi.org/10.1177/0734242X14565543.

Lakhan, C. (2015). Evaluating the effects of unit based waste disposal schemes on the collection of household recyclables in Ontario, Canada. *Resources, Conservation and Recycling*, 95, 38–45. https://doi.org/10.1016/j.resconrec.2014.12.005.

Manitoba Conservation and Climate. (2021). *Manitoba waste diversion and recycling framework review.* https://gov.mb.ca/asset_library/engagemb/climate/wastediversion/recycling-framework-executive-summary.pdf.

Multi-material Stewardship Manitoba. (n.d.). *2020 annual report: Celebrating 10 years.* https://stewardshipmanitoba.org/wp-content/uploads/2021/06/Web-File-MMSM-Annual-Report-2020.pdf.

Sebastian, R. M., & Louis, J. (2022). Waste management in northwest territories, Canada: Current practices, opportunities, and challenges. *Journal of Environmental Chemical Engineering*, 10(1), 106930. https://doi.org/10.1016/j.jece.2021.106930.

Solid Waste Association of North America. (2024). *SWANA reports increase in 2022 worker fatalities.* https://swana.org/news/swana-news-archive/article/2023/03/06/swana-reports-increase-in-2022-worker-fatalities#:~:text=For%202022%2C%2046%20solid%20waste,%2C%20SWANA%20Executive%20Director%20%26%20CEO.

Statistics Canada. (2021). *Waste management industry: Financial information, 2018.* www150.statcan.gc.ca/n1/en/daily-quotidien/211209/dq211209d-eng.pdf?st=hWfWSu3P.

Statistics Canada. (2022a). *Canada's core public infrastructure survey: Solid waste and asset management, 2020.* www150.statcan.gc.ca/n1/en/daily-quotidien/221028/dq221028b-eng.pdf?st=VPZK8NSC.

Statistics Canada. (2022b). *The dirt on composting.* www.statcan.gc.ca/o1/en/plus/1327-dirt-composting.

Statistics Canada. (2023). *Census profile, 2021 census of population.* www12.statcan.gc.ca/census-recensement/2021/dp-pd/prof/search-recherche/lst/results-resultats.cfm?Lang=E&GEOCODE=46.

Statistics Canada. (2024). *Biennial waste management survey: Waste diversion, 2022.* www150.statcan.gc.ca/n1/daily-quotidien/240408/dq240408b-eng.htm.

United Nations Environment Programme. (2023). *Towards zero waste: A tool for delivering the Sustainable Development Goals.* https://www.unep.org/resources/report/towards-zero-waste-catalyst-delivering-sustainable-development-goals.

United Nations Environment Programme. (2024). *Global waste management outlook 2024: Beyond an age of waste: Turning rubbish into a resource.* https://wedocs.unep.org/bitstream/handle/20.500.11822/44939/global_waste_management_outlook_2024.pdf?sequence=3.

Yin, R. K. (2018). *Case study research and applications: Design and methods* (6th ed.). Sage.

7 Municipal solid waste management in First Nations

Examining operational challenges and solutions

Anderson Assuah

Introduction

Effective management is required to deal with the increasing amount of municipal solid waste (MSW) that is being generated globally. Of the 2.24 billion tonnes of MSW generated in 2020, about 38% were identified to be uncontrolled waste—waste not collected, and or collected, but dumped or burned openly (United Nations Environment Programme, 2024). While there is great need to focus on upstream management efforts (e.g., prevention and reduction) to conserve resources (Herbst & Barner, 2024), downstream management such as disposal continues to dominate MSW management approaches in many countries such as Canada. In Canada, approximately 97% of waste meant for disposal is landfilled and 3% is incinerated (Government of Canada, 2024). The amount of waste incinerated or disposed of in landfills increased by 7% from 2002 to 2018 (ECCC, 2022) and by 8% from 2002 to 2020 to reach 26.1 million tonnes (ECCC, 2023). These statistics show a preference for waste disposal.

Given that most of Canada's MSW is disposed of through incineration and or landfilling (downstream management), management at this level requires a lot of attention by those tasked with the day-to-day management of waste. This makes the role played by MSW employees such as managers and operators essential because they need to have knowledge and understanding of the systems they operate, adequate training, resources, and the know-how to properly operate equipment (Deress et al., 2019). For instance, managers who understand the recycling policy of their provinces and municipalities are better placed to prevent or reduce recycling contamination than those who have less knowledge of acceptable recycling practices.

However, many First Nations communities in Canada are at a disadvantage because they struggle to adequately manage MSW. This is because communities generally lack the needed infrastructure, facilities, programs, and resources (Oyegunle & Thompson, 2018; Oceans North, 2021) to provide comprehensive MSW services to community residents. These factors make managing MSW difficult for communities and managers, who often must work with little resources.

DOI: 10.4324/9781003534129-10

Consequently, this chapter examines the operational challenges that First Nations MSW managers and operators face and their efforts to enhance existing programs, manage facilities, and deal with challenges.

Methods

This was a qualitative, multiple case study that gained comprehensive understanding of MSW management in 12 First Nations communities (Yin, 2018; Gustafsson, 2017). We focused on First Nations south of the 60th parallel, and searched for cases in western Canadian provinces, including Alberta, British Columbia, Manitoba, and Saskatchewan, because they collectively have a higher First Nations population according to data from Statistics Canada (2023).

Given that there are many First Nations in each province and there is no publicly available data on their MSW management programs, we looked to see which First Nations were embarking on initiatives to improve their existing MSW management systems. We did so by speaking to employees of Indigenous Services Canada (ISC) that work directly with First Nations on MSW management. Specifically, we asked questions about communities that had the existence of recycling, reuse, and composting systems and programs; operated an eco-depot or transfer station, and/or a landfill/dumpsite; and sorted their waste. First Nations communities that met the above criteria were selected.

Based on interviews with six ISC employees, 20 communities were recommended, out of which 16 met our criteria. Three of the 16 communities declined to participate in the research, and we were unable to reach and engage with anyone in one of the communities despite our best efforts. This process resulted in 12 communities that participated in the research. There were three communities each from the four provinces. Only one community out of the 12 did not have year-round road access.

After identifying the communities, MSW managers and operators in each community was contacted and interviewed because of their involvement in daily operations. The interviews lasted for about an hour each, and the questions generally focused on MSW management systems and programs, including operational challenges and opportunities. The interviews were transcribed, and transcripts were sent to all participants for member checking. The resulting data was organized using NVivo (QSR, 2010) software, and themes grounded in the data were derived. These themes are discussed below using pseudonyms; for example, P1, P2. This was done to cater for anonymity and confidentiality of participants, as agreed to in the research's ethics. Likewise, individual community names are not mentioned or associated with the data. However, quotations are used extensively to highlight information provided by participants. This research was conducted in line with the Ownership, Control, Access, and Possession principle for conducting research with First Nations. Thus, each community that participated in the research approved its participation.

Results and discussion

Operational challenges

Unsorted or unsegregated waste

A common challenge that all participants mentioned was that many community residents do not source-separate their waste before bringing them to the facility or struggle to place waste streams in their designated areas at their facilities. According to P2, this was a major challenge, particularly when they started their community's recycling program.

> When we started operations, segregation was a challenge, because people just mixed their garbage. So, we had to tell people to put electronics here, cardboard there, etc. That was the obstacle when we started. People had it in their minds that all garbage is garbage, but they gradually got to understand it.

P5 added: "They dump everything in the bin, they do not sort it out. We sort it out … I send monthly letters to community members that they should sort out their waste before dumping them, but some do not do it." For some participants, this is a continuous challenge that they must deal with daily. P3, for instance, described their encounter with a community member.

> A lot of people still do not recycle, so they put everything together. We are open 6 days per week, but some still drop off their garbage after closing hours. I have to keep reminding people about it. Some people are just headstrong about it, but I must keep doing it. That is the down part of it: people have poor attitude about recycling. Everywhere you go, you will find people who have poor attitudes, they do not like to be told what to do, and they do not want to hear the truth! For instance, there is a guy who has been here about four times, and I tell him, "You have to recycle, buddy," and he cusses at me. And I tell him, "You are not dropping all this garbage here, you have to recycle them." He has a whole truck load of garbage. He tells me that if I am not going to allow him to drop them off, he is going to the mall to drop them off in front of the Chief's store. I told him oaky; I will let them know that you are coming.

P4 summed up this challenge up as follows:

> Getting people to deal with their recyclables, composting, and reducing household garbage is the biggest challenge. I should be shipping garbage out in three weeks, not two or 10 days. My household bin and recycle bin should be filling up at the same time. So, getting the information to people to do all these is the challenge.

Source separation of MSW is important to the efficiency and success of MSW management programs, particularly waste diversion. Accordingly, communities need to provide clear information and demonstrate proper sorting practices (e.g., via videos) to residents to achieve this goal. This is essential because residents can get confused about which bins their waste should go into and thereby likely to dispose of them in the garbage or throw them in recycling bins, where they do not belong. Specifically, to recycling, a poor source separation system can lead to contamination, which reduces the recycling rate of programs (Chung, 2018; Smith et al., 2018).

Unaccepted waste streams

Waste facilities in the communities have waste streams that are accepted and those that they are incapable of dealing with. However, this has not stopped community residents and sometimes people from neighboring communities from accessing and illegally disposing of unaccepted waste streams. P6 revealed, "Some of the contractors we deal with here dump stuff at places where they are not supposed to dump them, and we are getting housing department to deal with them." According to P2, their community is dealing with a lot of unacceptable waste streams.

> The hardest [unacceptable waste streams] we are having is construction waste. Some contractors haul their waste to the waste transfer station, and we tell them that we do not take those, because in their contract, they are supposed to haul them to the regional landfill and weigh them and get a receipt, which the Band pays for.

The above quotations show a lack of understanding of the waste management programs that the communities are embarking on and or a reflection of inadequate stakeholder involvement in planning the programs. Waste management stakeholders are diverse and their ideas and involvement in the planning and management process can be vital to the success of programs (Le et al., 2018). If contractors are disposing of their waste at these existing facilities, it is likely they do not understand the system or do not have enough information regarding how the programs work. Alternatively, it could also be an attitudinal and behavioral problem by the contractors, who can be fined to ensure they comply with the rules.

For their part, P12 explained how members from neighboring communities illegally dispose of their waste at their facility:

> Some of our neighbouring communities sneak in at night and dump the garbage behind our gate. Some of them hide their number plates, and they come in when they know our guys are not operating on a particular day. Most of the things are used furniture. What we are trying to do is to get a camera installed, so we can catch the people doing it. Some people have old vehicles in the community that can be taken away, but we do not take that at the waste transfer station, so that will be a different business. A lot of people are hunters, and we do not take the skin, hide, or bones from animals.

Limited financial resources

All participants agreed that a major challenge to their operations is finances or the costs they incur to manage their facility and programs. P3 explained this in detail about their recycling program.

> Nothing is free, everything costs. Cost is the challenge to the program, and operations but it is worth it. We are in the hole every year. People think we make money out of this. One guy came in and said he wanted a metal, and I asked what he wants specifically. He said he wanted metal to go sell to make money. I told him "No" at least five times, but he would not take that for an answer. He started telling us we are just making money here and we do not want to give him the metal. I told him we do not make money here; rather, we spend a lot running the program, but we are always in a hole each year. I am not doing this to make any money.

Majority of participants explained that they do not receive enough money from ISC for their operations and that they hardly get funded by Chief and Council. This limited or lack of funding for programs and maintenance of facilities negatively impacts how they operate. To change fortunes and encourage more funding for operations, P11 explained, "The next challenge is the cost involved—making people understand that managing the waste transfer station cannot be done on a shoe-stringed budget, but you need to invest a lot of money into it for it to work properly."

Funding continues to be a major challenge to the efforts by many First Nations communities to improve their MSW management systems (Oyegunle & Thompson, 2018). Although the First Nations Solid Waste Minimization Initiative was started in 2016–17 to respond to this age-old need for solid waste funding, funding is still a challenge for many, especially remote and rural First Nations (Indigenous Services Canada, 2021). Like municipalities, some First Nations communities participate in stewardship programs, specialized recycling initiatives, or find ways to sell recyclables to generate extra funding to reduce the cost of operations. The effects and impacts of insufficient funding are explained below.

Equipment breakdown, infrastructure, and facilities

Some communities are unable to purchase new equipment for operations and/or repair broken down equipment because of inadequate funding. This has added to the operational issues the communities face in their quest to improve programs and manage their facilities. P9 recounted that in their community

> we do not have a garbage truck to collect the garbage; rather, we use the Band's half-tonne garbage truck to do the collection. We do not have equipment to process the garbage as well. We do not also have capacity to sort out recyclables like bottles and cans.

Similarly, P3 reported:

> The problem with the landfill or dumpsite is that we do not have a caterpillar or our own bulldozer to push up and flatten the landfill/dumpsites. The community has one, but it is being used and we have not had it for about 2 months. We have our own John Deere skid steer, but it is not big enough. Not heavy enough to compact the dumpsite. So, one main challenge is not having our own bulldozer.

These quotations further highlight challenges with funding discussed above and may also reflect inadequate maintenance planning. This is because MSW management is a capital-intensive undertaking and requires consideration for needed equipment, such as bulldozers and trucks prior to implementing programs. In fact, the above equipment and others should be secured before any form of operations is started.

Two participants mentioned that they have inadequate land to expand their operations. "The challenge is that we need to have a large land for our facility, because we are limited in terms of land," said P8. According to P10:

> We have a limited land base because we are basically in a swamp, so we do not have a good land area to manage the waste transfer station. I think ISC's calculation of the amount of household garbage to be picked up was not accurate and that has contributed to the failure of the waste transfer station.

Locating stewardship organizations

An important part of managing MSW waste is participating in extended producer responsibility or producer responsibility programs, so that recyclable materials collected can be properly dealt with. Stewardship organizations or producer responsibility organizations are tasked with the responsibility of dealing with the end-of-life of products (Tasnim et al., 2024). Although there are stewardship programs and organizations in all provinces that First Nations communities can participate, some participants mentioned that it is difficult to locate these organizations while others do not participate in the programs. This has negatively impacted their operations, because collected recyclable materials do not only occupy space in facilities but take considerable amount of financing, logistics, and planning to properly dispose of, especially in rural and remote First Nations communities. P11, for example, explained:

> The one [waste steam] that we have challenges with is the tires. We have 4,500 tires in various shapes and having a hard time to get rid of them. We have them stacked up, so they do not cause harm to the environment. It is probably going to cost us about $30,000 to $35, 000 to get rid of the tires. I talked to some people, and I told them that if I do not have to pay to put them in their facility,

> I am willing to put them in a semi-trailer to dispose of them. That is the biggest environmental issue we have here.

Rural and remote First Nations communities find it challenging to participate in various stewardship programs because of several factors, such as lack of infrastructure, complex paperwork, and long distances from recycling facilities (Assuah & Sinclair, 2021; Oceans North, 2021). Working with many individual stewardship organizations without dedicated staff to complete often tedious paperwork presents a challenge to communities. To deal with this phenomenon, waste stewardship organizations in some provinces are starting to work together under one "umbrella" to provide services for First Nations communities. Examples are the Backhaul Program in Manitoba (Green Action Centre, n.d.) and the First Nations Recycling Initiative in British Columbia (Recycle BC, n.d.) that provide support to communities to divert waste from their landfills or dumpsites. These programs save First Nations time and resources looking for individual stewardship organizations or private recyclers to work with.

Efforts to enhance operations and management

Continuous education and awareness creation in the community

Participants agreed that successfully managing their programs and facilities, and positively changing attitudes and behaviors require that community members are continually informed and educated about MSW management. As a result, they have adopted various measures and strategies, such distributing pamphlets, onsite teaching and demonstrations, radio, and word of mouth to achieve this. These measures are generally geared toward making "sure people follow the rules, because when they slip up once, twice, things go south" (P11). Education and awareness creation has occurred at different stages of management. For example, at the outset of their programs, P2 revealed:

> We worked 7 days a week and 14 hours a day when we first started, so that we could educate people. We did that for a year. Then we studied the days and hours to see which days people come in to dump their garbage and the time. As a result, we shortened our hours from 14 hours to 8 hours a day to operate from 9:30am to 6pm, Sunday to Thursday (5 days a week).

MSW management requires initial and ongoing learning, education, and information sharing if peoples' attitudes and behaviors are to positively change. As a result, employing a variety of ways to get information out, as participants mentioned, is important. Given the importance of culture to First Nations, utilizing Indigenous ways of learning, such as ceremonies, Elders/Knowledge Keepers, and storytelling (Nepinak, 2019) can be culturally appropriate ways for community residents to learn about MSW management and participate in programs. The overarching goal of these methods should be to make residents (re)connect with the environment and assume responsibility for its protection.

Being friendly and making community residents who visit the facilities feel at home was a strategy used by other participants, such as P6, who mentioned, "We educate our tribal people that 'hey, this is my place, and I have to take care of it!' They have to feel comfortable when they come in here." P3 further narrated how they go about this:

> When people come in here, they know where to put stuff and, if they do not, we tell them where to put them. If there are lots to do, the other co-workers help. Some people listen, and others do not. You have to keep talking to people, explaining to them why it is important. No matter who you are, you like some compliments, so when people drop off their recyclables when I am working in the shed, I say to them, "Right on! Thank you. Have a good day!" That is how I relate with everybody … That works on people, having a happy attitude at the workplace. There is still going to be negative people who do not listen.
>
> All participants highlighted the need for constant, consistent, and regular communication about their programs and operations in their efforts to educate and inform community members.

As highlighted by P11,

> Communication is key because you can have the infrastructure, but without good communication it will fail. For the people, it is how best to spin it, so that they can understand the system. The challenge therefore lies in communicating well and properly with people regarding what is being done. We put information on our radio station, and now they understand because we put a lot of information on there. They now fairly understand the system, so they go there and do what they need to do and get out of the place.

Learning from community residents and programs

Learning from their own programs and feedback or complaints of community members is one of the ways participants have been dealing with the challenges and barriers they face. P12, for example, indicated that their community is learning about "The steps we have to take to collect the garbage and also sending it to the proper facility instead of piling up here." P6 also added:

> We are learning from the community. For instance, the population is increasing, and we started with three large bins, but we now have eight of them. We are realising that contractors need bins, our department needs about two, and the community also needs some of the bins. We are learning as we go.

On learning from community complaints, P7 indicated, "They complain we are not picking up enough or often, so we learn from their complaints." For their community, P11 reported that they are

> learning the way the community thinks, how they see things. You explain to them and make the explanations realistic. Once they understand you, they will help you. A lot of them also have good advice, so you listen to them; it is not a one-way street, a lot of them give me advice, I research and try to implement the good ones.

Learning from the people for whom programs are being implemented is an avenue to gauge the health of a program and make necessary adjustments. As stakeholders, community residents can be helpful in identifying components of programs that are effective and challenges that waste managers may overlook or unable to identify and deal with. Therefore, actively involving stakeholders in managing the program, such as listening to their concerns and ideas, can be a catalyst for success (Helmefalk et al., 2023), because they feel involved, heard, and a sense of contributing toward a common community goal.

Learning from the operations of other communities

To properly operate and deal with the challenges they face, some participants mentioned how learning from other communities—near and far—has been helpful. Visiting landfills and waste transfer stations of other communities, participants explained the impact of their learning. According to P2:

> We went out to three municipalities and asked them how they deal with their waste transfer stations. Some were good, and others had challenges, so rather than reinventing the wheel, we took the best practices. Some had push pits, non-compactors, which were loose etc. Since we are in a very windy part of the province, we saw that compactors were the best for us here, so that we do not get the wind blowing them away. One thing we learned is that when your population is on your side, participation starts increasing. There is initial resistance, but once they know the positive impact, they will get behind what you are doing—that is one thing that we learned from the municipalities.

For their part, P1 explained how they learned from their neighbors:

> We have the [name withheld] people as our neighbours … One thing we have learned from them is the Free Store [Reuse Store]. We were researching a lot on that, and we looked at how other communities are running their Free Store, and I think we have achieved success. Some people just like to get rid of it rather than repurposing it for other people to use. We did not want people hounding around and looking into piles. We have also learned about regulations. When we go for our training, which we have done with SWANA, we go to other

> communities to learn about how these are run, so we see them in operation. Lots of our guys have gone to different landfills and eco depots to see how well they are controlled there.

P12 added:

> We looked at [name withheld], which was already operating their waste transfer station and were ahead of us. The operator walked us through what they do daily and that was helpful for us and our guys. We learned about how they were collecting their oil and paint, but we had no capabilities yet.

By learning from other communities with similar systems and demography, the communities prepared themselves for what they should expect, which would have expanded their knowledge base before starting their own programs. Additionally, common pitfalls were avoided, as indicated by participants. Building relationships with other communities, particularly neighboring communities, can also be an avenue to collaborate, create partnerships, and take advantage of economies of scale.

Networking

Some participants explained that connecting and building relationships with other managers and operators at workshops and conferences have increased their knowledge about operations, which they have applied. For example, P3 emphasized that networking has helped them to locate provincial stewardship organizations:

> Yes, workshops are very helpful. My co-worker was with me, and he was in one room, and I was in another. I took as much notes as I could. I guess they call it networking. I took notes on presentations on oil, bins, etc. The oil guys, for instance, came in here, and hauled lots of oil away. Basically contacts, who to contact. Getting in touch with the White Goods [e.g., fridge and microwave] guys, Multi-Material Stewardship Manitoba, Tire Stewardship, etc. was just networking and getting their phone numbers. I write a log on what they do when I take their phone numbers. That helps me to phone someone when I need something.

P5 has utilized the knowledge gained about landfill operations in their community. They explained:

> I have learned from the workshops, and I explain to the guys that work at the facility. For example, I tell them to cover the landfill and sort out the steel from the garbage. We get the books and re-read them and see what we must do.

In addition, P8 explained that they are "familiar with the Coast Waste Management Association Board. We have an annual conference and some events in the year inviting some speakers to come in to talk. By getting involved with that, I learn from the presentations and talks."

Networking opportunities, such as conferences provide platforms to learn about new management methods and approaches, share experiences, and connect with those in the waste management sector. This could be potentially beneficial to First Nations communities, who often have smaller populations and limited resources to provide comprehensive MSW management services. This was the case for the participant whose community benefited because of the connections they made with producer responsibility organizations in their province. Thus, First Nations communities, through networking, can take advantage of services and resources provided within these networks.

Conclusion

This research examined the operational challenges that First Nations MSW managers and operators face in their work and efforts to enhance management of programs and facilities. Attitudinal and behavioral challenges of community residents was a major challenge that affects daily operations. Illegal dumping of construction waste by contractors and residents of neighboring communities and limited finances were other challenges that were revealed. Nevertheless, the managers and operators are embarking on community education to change attitudes, learning from community residents and other communities, and taking advantage of networking opportunities.

These findings show that consistent and ongoing education of community members, and strong communication between MSW managers and community residents are important in MSW management operations because residents feel part of managing the programs. Strong and effective communication is needed and even more essential in First Nations communities, in which people know each other because of the communal way of living. It is therefore pertinent for managers and operators to build personal relationships with community residents in their work. It is recommended that managers incorporate Indigenous ways of learning, such as storytelling and ceremonies when developing community education programs and campaigns to enable residents (re)connect with the land and protect it.

Furthermore, overcoming MSW operational challenges requires learning, networking, and taking advantage of these networks, which some managers and operators are pursuing. These avenues have broadened and deepened their understanding of their roles and responsibilities. Continuous and regular education is therefore required of managers, so they can share and pass down best practices to community residents.

Finally, while operational challenges exist, providing the needed support and resources, coupled with adequate MSW planning, can improve MSW and smoothen operations. That is to say that communities that view poor MSW management as a common concern and are actively working to improve their current situations need to be supported by solid waste stakeholders. For example, provincial governments can supplement federal funding received by communities and provide best practices resources/guides to them. As demonstrated by some waste stewardship groups in Manitoba and British Columbia, working under a single project to provide services to First Nations communities encourages them to participate in stewardship programs, and saves them time and resources searching for individual stewardship organizations.

This research adds to the scanty body of literature on First Nations MSW management and mainstreams the experiences of MSW managers and operators. Additionally, it offers practical strategies that other First Nations and rural and remote communities can implement to improve their operations and management. Future research can examine the occupational hazards that MSW managers and operators experience and strategies to overcome them.

References

Assuah, A., & Sinclair, A. J. (2021). Solid waste management in western Canadian First Nations. *Waste Management*, 129, 54–61.

Chung, E. (2018, April 6). *Many Canadians are recycling wrong, and it's costing us millions.* CBC News. www.cbc.ca/news/science/recycling-contamination-1.4606893.

Deress, T., Jemal, M., Girma, M., & Adane, K. (2019). Knowledge, attitude, and practice of waste handlers about medical waste management in Debre Markos town healthcare facilities, northwest Ethiopia. *BMC Research Notes*, 12, 1–7.

Environment and Climate Change Canada. (2022). *Canadian environmental sustainability indicators: Solid waste diversion and disposal.* www.canada.ca/en/environment-climate-change/services/environmental-indicators/solid-waste-diversion-disposal.html.

Government of Canada. (2024, June 21). *Municipal solid waste management in Canada.* www.canada.ca/en/environment-climate-change/services/managing-reducing-waste/municipal-solid/environment.html.

Green Action Centre (n.d.). *2023-2024 Impact Report.* https://greenactioncentre.ca/wp-content/uploads/2025/01/Impact-Report-23-24.pdf

Gustafsson, J. (2017). Single case studies vs. multiple case studies: A comparative study. (Unpublished manuscript). Academy of Business, Engineering and Science, Halmstad University. Retrieved from www.diva-portal.org/smash/get/diva2:1064378/FULLTEXT01.pdf

Helmefalk, M., Palmquist, A., & Rosenlund, J. (2023). Understanding the mechanisms of household and stakeholder engagement in a recycling ecosystem: The SDL perspective. *Waste Management*, 160, 1–11.

Herbst, J. M., & Barner, L. (2024). Waste to resource recovery at a marina: Empirical evidence of upstream and downstream innovation for circularity. *Journal of Environmental Management*, 359, 120942

Indigenous Services Canada. (2021). *Evaluation of the First Nations solid waste management initiative.* www.sac-isc.gc.ca/eng/1640350964188/1640351002909.

Le, N. P., Nguyen, T. T. P., & Zhu, D. (2018). Understanding the stakeholders' involvement in utilizing municipal solid waste in agriculture through composting: A case study of Hanoi, Vietnam. *Sustainability*, 10(7), 2314.

Nepinak, S. (2019). Exploring our roots: Knowledge gathering with the elders. https://irbu.arcabc.ca/islandora/object/irbu%3A444/

Oceans North. (2021). *Towards a waste-free Arctic.* https://oceansnorth.org/2021/03/23/towards-a-waste-free-arctic-new-report-highlights-urgent-need-for-better-waste-management/.

Oyegunle, A., & Thompson, S. (2018). Wasting Indigenous communities: A case study with garden hill and Wasagamack First Nations in Northern Manitoba, Canada. *Journal of Solid Waste Technology and Management*, 44(3), 232–247. https://doi.org/10.5276/JSWTM.2018.232.

Recycle BC. (n.d.). *First Nations recycling initiative.* https://recyclebc.ca/projects/first-nations-recycling-initiative.

Smith, L., Tam, J., Martel, P., & Holleran, J. (2018). Waste management at Dalhousie University, Canada: Social factors affecting recycling behavior and use of the four bin system at the Studley campus. https://dalspace.library.dal.ca/items/30fa9109-7c18-4e09-9591-7dd67240d99d

Statistics Canada. (2023, October 25). *An update on the socio-economic gaps between Indigenous peoples and the non-Indigenous population in Canada: Highlights from the 2021 Census—A compendium report to the department's 2023 annual report to parliament.* www.sac-isc.gc.ca/eng/1690909773300/1690909797208.

Tasnim, A., Chowdhury, R., Mim, S. J., Ng, K. T. W., & Adu-Darko, H. (2024). Influence of Canadian provincial stewardship model attributes on the cost effectiveness of e-waste management. *Journal of Environmental Management*, 358, 120945.

United Nations Environment Programme. (2024). *Global waste management outlook 2024: Beyond an age of waste—Turning rubbish into a resource.* https://wedocs.unep.org/20.500.11822/44939.

Yin, R. K. (2018). *Case study research and applications: Design and methods* (6th ed.). Thousand Oaks, CA: Sage Publications.

Part 3

Community-based waste management approaches

8 Contribution of waste pickers' grassroots organizations to waste management in Victoria and Vancouver

Ana Maria Rodrigues Costa de Castro and Jutta Gutberlet

Introduction

In the face of escalating global waste challenges and the urgent call for circularity in resource management, the issue of waste generation and its efficient handling has become paramount. Globally municipal solid waste generation alone is predicted to increase another 56% from 2.1 billion metric tons in 2023 up to 3.8 billion metric tons by 2050 (ISWA, 2024). This growth pattern also extends to Canada, where the data shows that between 2002 and 2020, the total amount of solid waste generated had increased 17% or by 5.3 million metric tons to reach 36.0 million metric tons in 2020 (Environment and Climate Change Canada, 2023). Despite notable progress in certain areas, such as municipal recycling programs and organic waste management initiatives, there remains substantial room for improvement. Issues persist in reducing waste generation, enhancing organic waste recovery, reducing the generation of plastic waste and improving the recycling rates of specific materials like plastics. The situation demands innovative solutions to promote recycling and reduce the waste of precious resources.

Urban political ecology (UPE) is a theoretical framework that critically examines the sociopolitical dynamics shaping urban environments, emphasizing the unequal distribution of resources and environmental burdens. At its core, this approach interrogates the interplay between power relations, ecological processes, and urban development (Bartels et al., 2020). Within this framework, solid waste management serves as a pivotal lens through which to analyze urban socio-environmental conflicts.

While the literature focuses on the link between solid waste and urban development in Global South contexts, the framework can also be addressed to situations in North American cities (Ernstson et al., 2021). Issues over how to deal with solid waste, whom to involve in waste management, or which technology to prioritize often reflect broader power struggles over land use, environmental justice, climate justice, and economic interests (Kalina, 2020). UPE underscores how these decisions are shaped by political processes, institutional arrangements, and the agency of various stakeholders, including communities, governments, and

DOI: 10.4324/9781003534129-12

corporations. By examining solid waste within UPE, scholars uncover how urban environments are not only physical spaces but also contested terrains where environmental inequalities and social injustices are reproduced and challenged.

Efforts like Extended Producer Responsibility (EPR) systems, including deposit refund mechanisms, aim to incentivize recycling and proper disposal. However, challenges persist, as many beverage containers end up in landfills instead of being returned through these systems, highlighting gaps in effectiveness and compliance (Phelan et al., 2022). Furthermore, only very few packaging materials are included under these schemes, gaps in the policies, due to differences in jurisdiction and lack of producer responsibility are still major bottlenecks of EPR (Diggle, 2022).

Waste pickers are essential actors, mostly known in the context of Global South cities, who operate within the informal recycling sector (Dias, 2016; Gutberlet, 2023). These socially and economically excluded workers also exist in the global North, however in smaller numbers, less organized and primarily specialized on beverage container recovery (Bulla et al., 2021; Porter, 2015). Yet, they play a critical role in reclaiming recyclables from public bins and streets, thereby bridging gaps left by formal waste management systems.

In response to these gaps, grassroots organizations have emerged globally, particularly in the Global South (Kain et al., 2022), advocating for waste pickers' rights and promoting their social inclusion and seeking to improve working conditions. Remarkably, while much research has focused on these organizations in lower-income countries, there remains a notable research gap regarding their role and impact in high-income nations like Canada.

This study seeks to address this gap by exploring the role of waste pickers in enhancing waste management and circularity in Canada, focusing specifically on two case studies in British Columbia (BC). By examining their contributions and challenges within the Canadian context, this research aims to shed light on how grassroots initiatives can drive sustainable practices and social equity in waste management systems.

Methods

The research is characterized as qualitative and exploratory, and each step is summarized in Figure 8.1. The methods consisted of interviews, field visits to one bottle depot, events and meetings, and analysis of secondary data carried out between October 2023 and June 2024, involving two waste pickers grassroots organizations. In addition to the data collected during this period, we also rely on the accumulated experience due to previous research in the field, since 2008, which contributes to the contextualization and discussion of our results.

The findings presented in this chapter are part of a larger research project with the North American Waste Pickers Alliance (NAWPA), which comprises five waste picker grassroots organizations, and the focus here will be on the organizations in BC, Canada: the Binners Project (BP) in Vancouver and the Diverters Foundation (DF) in Victoria. The terms binners and diverters are used for waste pickers in each city.

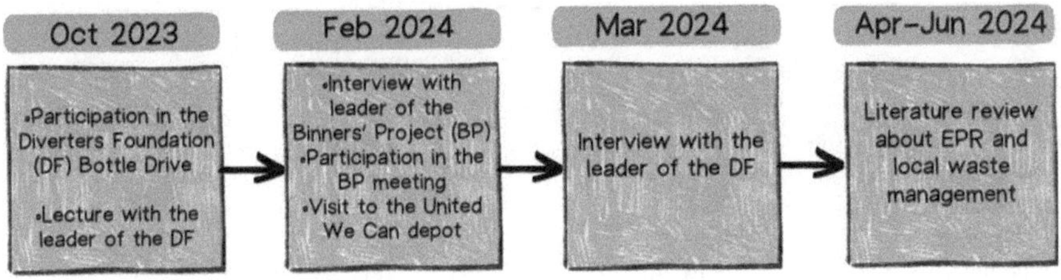

Figure 8.1 Research steps.

Source: Developed by authors.

The interviews were conducted with each leader of the two grassroots organizations. They were identified based on contacts one of the researchers has had with the groups for years and the invitations were sent by email. A semi-structured interview script previously approved by the Ethics Committee of the University of São Paulo, Brazil (code 76555923.2.0000.5390) was used, and the interviews were conducted, recorded and transcribed using Zoom. The data were then analyzed qualitatively using the MAXQDA program. Some of the information collected included reference to how the group was created, how many waste pickers participate, the activities carried out and how they promote the integration of waste pickers and contribute to the circular economy.

Additional data was collected in Victoria at an event called Bottle Drive organized by the DF and at a talk given by the group leader at the University of Victoria, both in 2023. In the following year a visit was made to a weekly meeting held with waste pickers from the BP in Vancouver and to the bottle depot that most of them use to deliver their containers. The objective of these visits was to get closer to the group, to learn more about their current work and to also talk informally with participating waste pickers, to gain insights into their perceptions of the initiatives, since we did not have the resources to go into the field to interview them. The additional data collected in the field was entered into MAXQDA for triangulation with the information from the interviews.

Study area

Victoria and Vancouver are both located in western Canada, in the province of BC. Vancouver is on the mainland, while Victoria, the capital of the province, is located in the south of Vancouver Island. Both cities bring natural beauty attracting tourism and are knowledge hubs, housing important universities and research centers. Table 8.1 provides some demographic indicators for both cities, also showing the level of vulnerability within different income groups.

According to the Core Housing Needs indicator, which measures the percentage of the population in each income category that lives in inappropriate conditions, a higher percentage of the very low and low-income population in Victoria is living

Table 8.1 Main characteristics of study area

City	*Vancouver (2021)*	*Victoria (2021)*
Population[a]	662,248	91,867
Population density per square kilometer[a]	5,749	4,722
Land area in square kilometers[a]	115.18	19.45
Percentage of households in Core Housing Need (CHN)* by monthly income category[b]	Very low income ($410): 71.61% Low income ($1,025): 61.35% Moderate income ($1,640): 35.24% Median income ($2,460): 4.94% High income (>$2,460): 0	Very low income ($337.5): 85.19% Low income ($843.75): 70.37% Moderate income ($1,350): 30.87% Median income ($2,025): 4.88% High income (>$2,025): 0

Sources: [a]Canada (2021); [b]Housing Needs Assessment (2021).

Note: *CHN means that housing does not meet one or more of the adequacy, suitability or affordability standards.

in inadequate conditions compared to Vancouver, indicating that poor people in Victoria have more housing issues (Housing Needs Assessment, 2021). A possible explanation for this data was mentioned in one of the interviews, which mentioned that Vancouver has more shelter spaces for people living on the streets. Another piece of information that can be extracted from this indicator is the value of each income bracket, which is higher in Vancouver and may be an indication that the cost of living in this city is higher than in Victoria. Yet, in both cities earning a livelihood is difficult for the poor, who suffer from financial and social vulnerability, projecting many people into homelessness and into informal recycling.

Results and discussion

Contextualization of waste management in the study region

In this section, we present relevant information about the management of municipal solid waste (MSW) in Victoria and Vancouver, to understand the system in which waste pickers operate and the existing power asymmetries that keep these informal actors marginalized. We include information on main policies, the actors involved in waste collection, and the amount of MSW generated, diverted and disposed of.

Waste policies

There are four decision-making levels: federal, province, regional districts, and cities. The policies of each of these levels most relevant to the scope of our study are presented in Figure 8.2. In 2009, the Canada-wide Action Plan for Extended

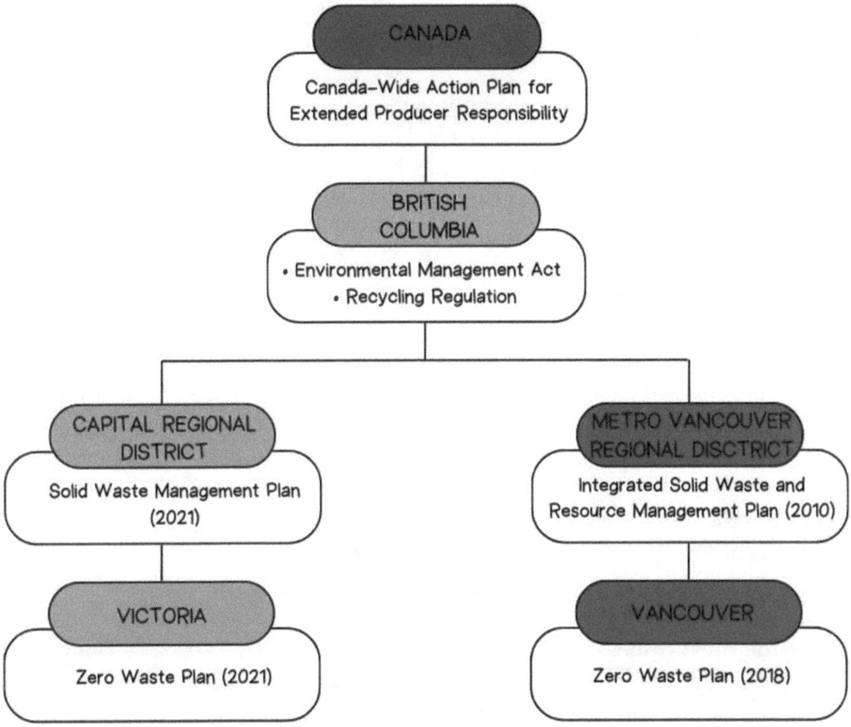

Figure 8.2 Policies related to waste in the study region.

Sources: Elaborated by the authors with information from Canada (2009); British Columbia (2003); British Columbia (2004); CRD (2021); City of Victoria (2021); Metro Vancouver (2010); City of Vancouver (2018).

Producer Responsibility was launched, shifting responsibility for the product life cycle from cities to producers (Canada, 2009).

Waste management in the province is regulated by the Environmental Management Act, which addresses management plans, disposal fees, littering, and greenhouse gases at waste management facilities, among other topics (British Columbia, 2003), and its Recycling Regulation which addresses EPR (British Columbia, 2004). BC already has EPR programs implemented for 18 categories of waste including: (a) beverage containers, (b) lead-acid batteries, (c) packaging and paper products; (d) paints, solvents, pesticides, gasoline, (e) pharmaceuticals; (f) tires; (g) used oil and antifreeze, (h) eleven subcategories of electronics and electrical (British Columbia, 2023).

The first EPR started in 1970 with a deposit refund system for bottles and cans, also known as Bottle Bill. In 2024, this system receives containers of all ready-to-drink beverages, except infant formula, meal replacements and dietary supplements, and pays 10¢ for each delivered item at the depots (Container Recycling Institute, 2023). This deposit refund system is specifically relevant to our research, as it is the main stimulus for binners in this region, collecting beverage containers for refund as main source of income.

Another relevant EPR to be highlighted is the one for packaging and paper products which was created in 2014. Under this system, the not-for-profit organization Recycle BC is responsible for curbside and multi-family recyclable collection and recycling depot services (Recycle BC, n.d.). The services are provided either directly (as in the case of Vancouver) or in partnership with governments (as in Victoria) or other organizations, which will be further explored in the next section.

The regional level is responsible to develop waste management plans. Victoria is part of the Capital Regional District (CRD) and Vancouver is part of the Metro Vancouver Regional District (MVRD) and both cities are subject to their regional waste plans. The CRD plan was last updated in 2021 and approved in 2023, while the MVRD plan was published in 2010 and approved in 2011. These documents establish important goals for waste management, such as reducing the amount generated (CRD, 2021; Metro Vancouver, 2010).

At a local level, the two cities created their zero waste plans, which seek to go beyond the recycling and composting policies that already exist in these locations, but move toward reducing waste generation, reuse, and maximum diversion of waste from final disposal, complying with the solid waste hierarchy. Both plans have specific actions for single-use items, construction and demolition waste, and organics (City of Victoria, 2021; City of Vancouver, 2018).

Only Vancouver seems to mention the informal sector contribution of binners, with the Metro Vancouver plan report mentioning support for the BP (Metro Vancouver, 2022); and the zero waste plan recognizing the project as an example of the city's zero waste initiative, in addition to having invited project representatives to workshops in preparation of the plan (City of Vancouver, 2018). Yet, as there is little to no access to resources and infrastructure for binners and diverters, they remain marginalized.

Waste collection stakeholders

The separation of MSW in the two cities occurs into three fractions, between compostable organics, recyclables (the Blue Box system), and "garbage" (which cannot be recycled or composted), and has different actors involved in the management. In Victoria and Vancouver, the city is responsible for collecting organics and garbage from homes, except multi-family buildings, where collection is carried out by private haulers, which also collect from industries, businesses, and institutions (City of Vancouver, 2018; City of Victoria, 2021).

The collection of recyclables is the responsibility of Recycle BC, under the EPR for packaging and paper, which has been the operator in Vancouver since 2016 (City of Vancouver, 2018), while in Victoria it continues to be carried out by the CRD, but financed by Recycle BC (City of Victoria, 2021). In addition to these actors are the diverters in Victoria, and the binners in Vancouver, collecting cans and bottles from the curbside Blue Box system and from public bins or litter in parks, beaches and streets. The city does not provide any support in these operations.

Waste quantification

Quantifying waste in the study region and comparing different levels (city, district, province, country) proved challenging, as there is no official database with year-by-year information. Therefore, the data found and displayed in this section (Figure 8.3) needs to be analyzed with caution, as multiple sources with different reporting periods were used to provide some insights on waste generation in the region.

It is estimated that Canada generates around 36,011,000 metric tons of MSW per year (Environment and Climate Change Canada, 2023), of which approximately 18% (6,586,061) is from BC (Lee et al., 2021). The amount of MSW generated in Victoria was equivalent to 1% of the total in BC (City of Victoria, 2021), while in

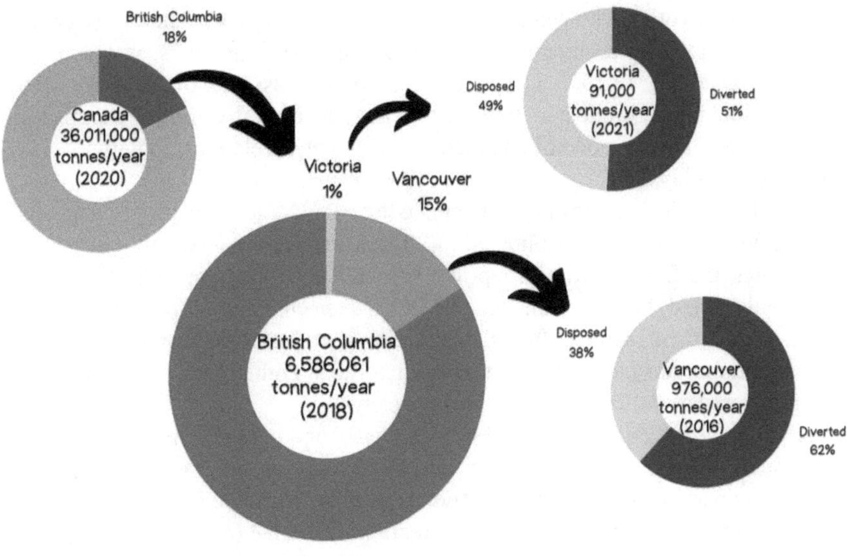

Figure 8.3 Amount of MSW generated in the study region.

Note: Total diverted includes the amount recycled and composted. Total disposed includes the amount sent to landfills or incinerators. MSW includes residential, commercial and construction waste in all cases, and in the Canadian amount also contains industrial waste.

Sources: Prepared by the authors with data from City of Vancouver (2018), City of Victoria (2021), Lee et al. (2021) and Environment and Climate Change Canada (2023).

Vancouver the value was around 15% (City of Vancouver, 2018), a bigger value as expected since the population is larger.

In both cities, more than half of the waste is diverted through recycling and composting (City of Vancouver, 2018; City of Victoria, 2021). The recovery rate was highest in Vancouver (62%), surpassing the average of 59% in BC (Lee et al., 2021) and 27% in Canada (Environment and Climate Change Canada, 2023). However, despite advances in terms of recovery, many wastes that could be recovered end up in landfills, as is the case of the CRD landfill, where 47% of the waste received in 2022 could have been diverted according to its plan (Capital Regional District, 2023).

Furthermore, the zero waste plans of both cities indicate that they still need to make progress in preventing and reducing waste generation (City of Vancouver, 2018; City of Victoria, 2021). These two challenges could be addressed by including the diverters and binners, as will be shown in the following section, which highlights the actions of these important environmental agents.

Grassroots waste pickers organizations

Group formation history

Figure 8.4 shows the timeline for the creation of grassroots recycling organizations in BC, which began in 1995 with the creation of the United We Can (UWC) bottle depot by Ken Lyotier, who was a binner himself. UWC is a social enterprise created to generate job opportunities and to mobilize support for binners, in Downtown Eastside Vancouver. In 2005, Lyotier created a binners association, in response to a new rule locking all dumpsters in the city. The binners started to use identification cards and developed a code of conduct, to seek a better relationship with the community (Tremblay, 2007).

In 2014, Lyotier's efforts began to create the BP in Vancouver, also with support from UWC, which was consolidated in 2015 when the organization was created. Years later, the BP created a group in Victoria, through a pilot project started in 2018 and named "Victoria Binners Project," later changed to The Diverters, to express their own identity. In 2020, with support from the Community Based Research Laboratory at the University of Victoria and other local actors, the group was registered as the DF, a not-for-profit society.

All three organizations, United We Can, the BP and the DF, continue to operate, however UWC and the BP currently no longer carry out joint actions. However, the binners participating in the BP activities usually return the collected containers to UWC. Due to its importance in the creation of the BP (and even other groups in Canada, not included in this chapter) and due to the reputation of being a depot where binners are welcome, a visit to the site was carried out to learn about the work and to talk informally with binners who use this place. It provides job opportunities and is a welcoming environment with daily opening hours, access to washrooms and where binners receive coffee and snacks every Wednesday.

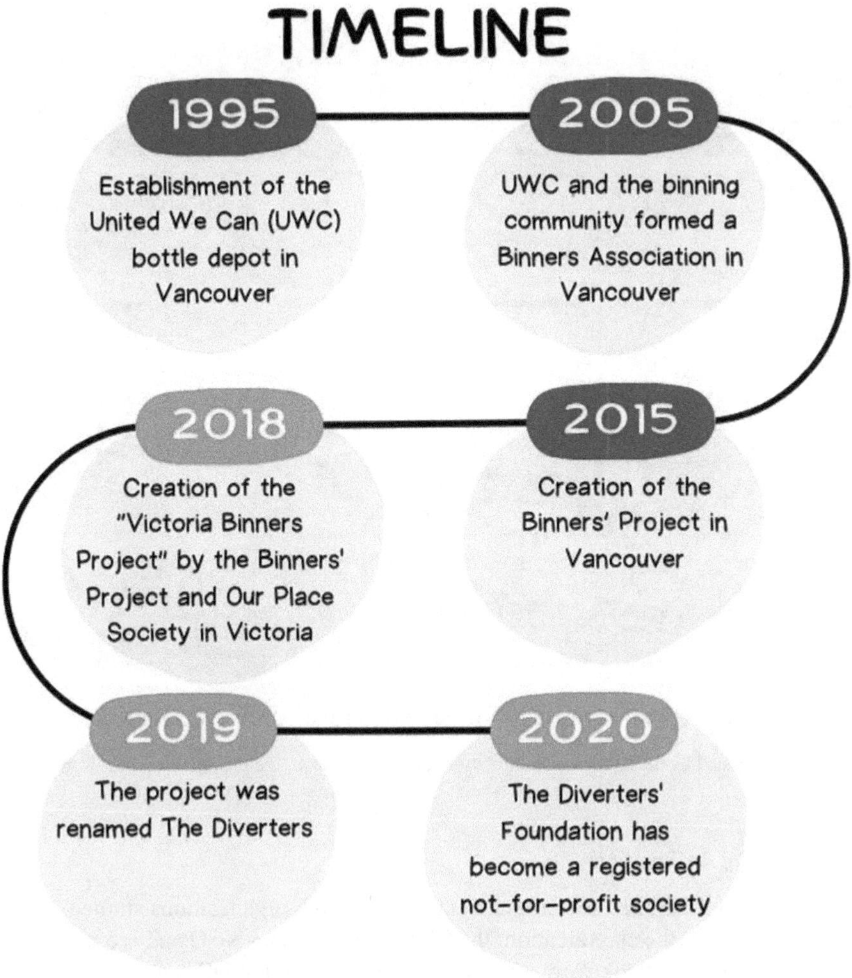

Figure 8.4 Timeline of grassroots organization creation.

Source: Created by the authors with information from Tremblay (2007) and the interviewees.

According to its website, the organization generates around 8,000 working shifts per year and have around 600 regular customers that are supported by its activities (United We Can, n.d.).

The BP and the DF continue to maintain contact and exchange experiences and both parties are part of the NAWPA, an initiative which helps destigmatize and value the informal activity in North America. Both organizations will be detailed in the next section.

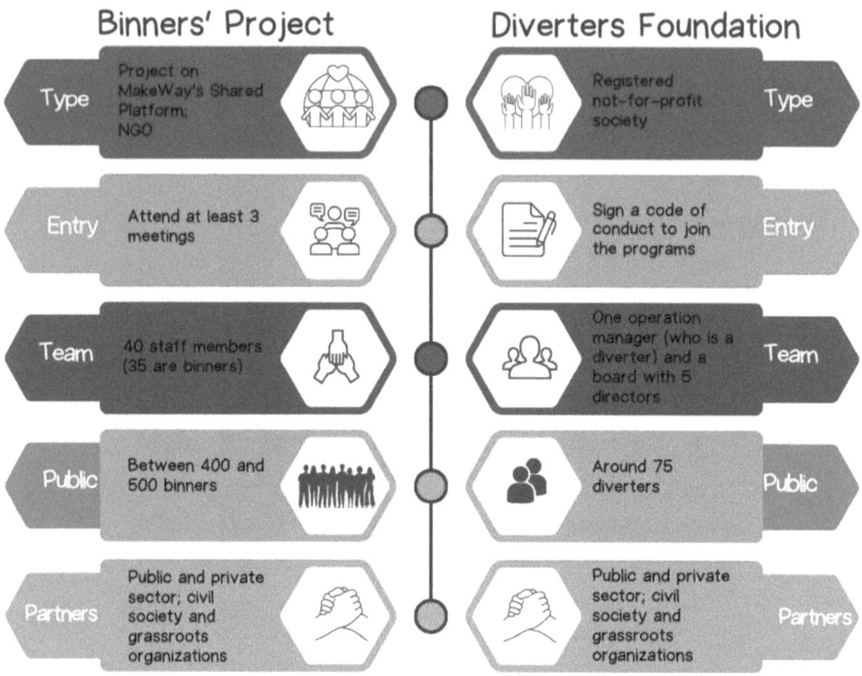

Figure 8.5 Main characteristics of the Binners Project and the Diverters Foundation.

Source: Elaborated by the authors with information from interviews and the organizations' websites.

Description of the BP and the DF

Figure 8.5 summarizes the main characteristics of the organizations studied. Their first difference is the classification: the BP is considered an NGO and is a project of the national charity MakeWay's Shared Platform, while the DF is a registered not-for-profit society. The key formal distinction is that not-for-profits serve the specific needs of a particular group or community and not the general public like NGOs do. To enter the organization, the member needs to attend at least three meetings in the case of the BP and sign a code of conduct for the DF, they do not need to pay any membership. In both cases, members carry out some activities together through organizations but continue to work independently collecting in the streets.

The BP is an older and bigger organization, currently with 40 staff members (35 of them are binners), and between 150 and 200 binners involved in activities. They estimate to have impacted almost 500 binners over the years. On the other hand, the DF is yet small scale, and has only one operation manager (who is a diverter) and a board of directors with five members who provide support. They estimate that currently 12–15 diverters are involved and that they must have impacted around 75 binners in Victoria since its beginning in 2020.

Regarding decision-making, in the BP the decisions related to the strategic direction or making changes are done with binners' participation, through consultations in small group meetings or votes during the weekly Tuesday meetings. These general meetings are a great opportunity for binners to meet each other, exchange information, and provide feedback to the project. For example, at the meeting that one of the researchers attended on February 13, 2024, there were 85 waste pickers present.

On the other hand, decisions in the DF are made by the operation manager, guided by the board of directors. It is important to highlight that the operation manager is a diverter who has lived on the streets and had his life changed thanks to the opportunity to work at the DF and, therefore, understands the reality faced by his fellows and can represent their voices. In order to become more representative, the DF has begun holding meetings with the diverters and other waste pickers who are not yet members. Finding an adequate meeting place seems to be a key challenge for this organization, since they do not have a headquarters.

Both organizations have partners from the public and private sectors and civil society organizations. Some examples are the cities of Victoria and Vancouver, the companies that hire for service provision, the University of Victoria which develops research projects, and other grassroots organizations such as the Cooperative Les Valoristes, from Montreal, with whom they exchange experiences.

The BP provides services, including back-of-house waste sorting in commercial and residential buildings, waste education and sorting at events, and awareness activities with the community including tours and public speaking. Besides that, the BP organizes training and weekly meetings with binners and has infrastructure projects with carts and hooks to facilitate the binners' work. the DF also provides hired services such as community and business refundable pick-up, sorting at events. There are also education and outreach programs, such as the Bottle Drive, which is an event in Victoria, in which returnable recyclables are received, and diverters can talk to the population. In 2024, two Bottle Drives have already been held, one in January and the other in March.

An activity that is common to both groups is the Coffee Cup Revolution, which started in Vancouver in 2014, replicated in Victoria in 2021, and takes place once a year. This event draws attention to the large quantities of discardable coffee cups, generated by the population and often incorrectly disposed of. On the day of the event, the organization pays 10¢ in refund per unit collected by the waste pickers (through funding received from partners). The event serves to simulate what would happen if this waste would be added to the EPR system, and how it could be recovered and simultaneously generate income for the workers. The collected cups are stacked in a public place to attract attention from the community. This is an advocacy and awareness-raising activity and also a call for action for the BC government (which is responsible for the EPR systems). In Vancouver, they already collected 610,000 coffee cups since the start of this Revolution (Binners' Project, n.d.).

Over the 9 years of work, the BP saw an increase in the number of binners participating in activities, and an increase in the number of clients interested in contracting services, resulting in more revenue and more job opportunities for binners. The interviewee also highlighted the positive outcome of having binners' voices heard by policymakers, who now consult the group in waste-related matters. Furthermore, the Coffee Cup Revolution has managed to recover more cups over the years, breaking a record in 2023 of almost 150,000 coffee cups collected in 3 hours during the event.

The DF highlighted as its main achievement the fact of having survived the pandemic, despite all the challenges of this period. Furthermore, they managed to create new projects over the years, diversifying their areas of activity. They also are consulted by the government on waste-related issues.

Despite the good results, both organizations face constant challenges, such as access to funding and fluctuation in the ability of members to participate and engage in activities. In summer, the weather conditions and the large number of tourists in both cities contribute to increase collection rates making waste picking work more attractive, according to the interviewees. In winter there is a reduction in activities and many waste pickers from Victoria migrate to Vancouver, where more shelters are available during the colder period, as reported by the DF manager, but we did not find records that show the exact numbers. Other challenges mentioned were the local political agenda often excluding the demands of binners and diverters, besides the lack of resources and members slowing the growth of the group, specifically in Victoria.

Main contribution to waste management and the circular economy

The literature points out the relevance of the work of waste pickers around the world for the circular economy (Zisopoulos et al., 2023; Gutberlet et al., 2017), combating plastic pollution (Cook et al., 2024), reducing greenhouse gas emissions and saving energy (Mesquita et al., 2023). Nonetheless, in this section, we will highlight the specific impacts identified in Victoria and Vancouver, summarized in Figure 8.6. It is noteworthy that all waste pickers can generate these impacts and contribute to sustainability and circularity. However, when they are organized, they can produce an even greater impact, which is why we highlight the importance of grassroots organization.

In summary, the work of the BP and the DF occurs where there are gaps in the municipal waste system, and when there is no competition with other service providers. These informal actors help solve persistent problems, such as littering in parks and beaches, or inadequate separation in public bins and at events. Therefore, these workers are contributing to waste management, bringing solutions to existing problems. Another valuable outcome is the creation of low barrier job and income opportunities for vulnerable populations.

The positive impacts highlighted in Figure 8.6 could be even greater if the refund system would also include other materials, such as coffee cups. The two groups have continuously demanded for the BC government to expand the EPR system, to include other materials, as explained above.

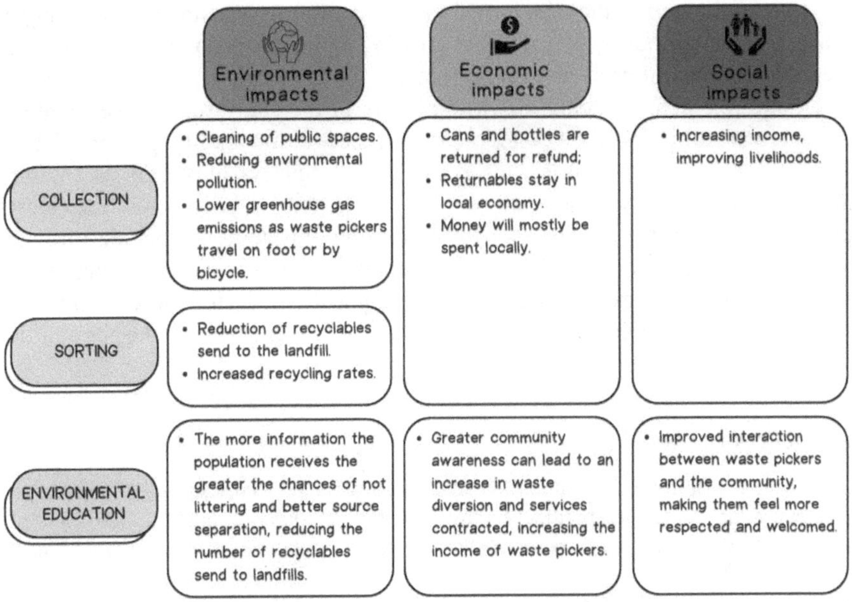

Figure 8.6 Services provided by local waste pickers and their impacts.

Source: Developed by the authors.

Conclusion

In this chapter, we showed two case studies of grassroots waste pickers in BC, highlighting their contributions to waste management, as well as a contextualization of the system and waste legislation in which they operate. The results showed that Vancouver and Victoria already have consolidated EPR systems and are investing in recycling and diversion of waste from landfills, but there are still gaps in their systems, which is where waste pickers act.

The UPE lens has helped understand the impact of marginalization and lack of participation of informal waste actors (binners, diverters). Existing policies and practices continue to prioritize standard approaches that exclude informal recyclers. While first signs indicate that the local government, specifically in Vancouver, has begun to acknowledge the contributions of binners, there is still great demand to further integrate their practices into formal systems, and to address their socioeconomic vulnerabilities within broader urban development strategies. We conclude that high generation of waste, littering and incorrect separation by the population are additional challenges.

The organization of waste pickers through the BP and the DF is important, as these groups promote their empowerment, training, and destigmatization, through the creation of job opportunities and advocacy for their rights. The BP is a more consolidated organization, with more employees and greater number of binners, and has contributed to the creation of Victoria's group. As a smaller and younger group, the DF still faces difficulties in growing, but it has been able to develop an increased number of projects over the years.

Despite the achievements of waste pickers in the study region, we highlight that there is still much to be done in terms of recognition and integration of these workers. We see opportunities, for example, in hiring waste pickers as environmental educators to help with waste reduction targets outlined in the zero waste plans. Furthermore, the expansion of EPR refund systems to include difficult-to-recycle waste such as coffee cups is an opportunity to generate income while increasing waste recovery. Finally, we highlight the need for more research on waste pickers in the Global North, to understand their demands and the possibilities to boost their integration.

References

Bartels, L. E., Bruns, A., & Simon, D. (2020). Towards situated analyses of uneven peri urbanisation: An (urban) political ecology perspective. *Antipode*, 52(5), 1237–1258.

Binners' Project. (n.d.). *Coffee cup revolution*. Retrieved July 2, 2024 from www.binnersproject.org/ccr.

British Columbia. (2003). Environmental Management Act. *SBC 2003 Chapter 53. Assented to October 23, 2003*. www.bclaws.gov.bc.ca/civix/document/id/complete/statreg/03053_00_multi.

British Columbia. (2004). Environmental Management Act. *Recycling Regulation. Deposited October 7, 2004. Last amended January 1, 2024 by B.C. Reg. 255/2023*. www.bclaws.gov.bc.ca/civix/document/id/complete/statreg/449_2004.

British Columbia. (2023). *Extended producer responsibility plans and annual reports*. Last updated on February 18, 2025. www2.gov.bc.ca/gov/content/environment/waste-management/recycling/extended-producer-responsibility/extended-producer-responsibility-reports-plans.

Bulla, J. P., Rendon, M., & Trenc, J. E. (2021). Policing the stigma in our waste: What we know about informal waste pickers in the Global North. *Local Environment*, 26(10), 1299–1312. https://doi.org/10.1080/13549839.2021.1974368.

Canada. (2009). *Canada-wide action plan for extended producer responsibility*. Canadian Council of Ministers of the Environment. https://ccme.ca/en/res/cap-epr_e.pdf.

Canada. (2021). *Census profile. 2021 census of population*. Catalogue no. 98-316-X2021001. www12.statcan.gc.ca/census-recensement/2021/dp-pd/prof/index.cfm?Lang=E.

Capital Regional District. (2021). 2021 solid waste management plan. *Approved by Ministry of Environment July 2023*. www.crd.bc.ca/service/waste-recycling/solid-waste-management/reports-publications.

Capital Regional District. (2023). *2023 solid waste management plan progress report*. www.crd.bc.ca/docs/default-source/crd-document-library/plans-reports/landfill-recycling/2023-reports/2023-solid-waste-management-plan-progress-report.pdf?sfvrsn=ff0dbdce_1].

City of Vancouver. (2018). Zero waste 2040. *Policy report*. https://vancouver.ca/green-vancouver/zero-waste-vancouver.aspx.

City of Victoria. (2021). *Zero Waste Victoria*. www.victoria.ca/home-property/waste-collection/waste-reduction.

Container Recycling Institute. (2023). Existing programs. *Last updated on August 28, 2023*. www.bottlebill.org/index.php/current-and-proposed-laws/canada/british-columbia.

Cook, E., Cano, N. S. de S. L., & Velis, C. A. (2024). Informal recycling sector contribution to plastic pollution mitigation: A systematic scoping review and quantitative analysis of prevalence and productivity. *Resources, Conservation & Recycling*, 6, 10758. https://doi.org/10.1016/j.resconrec.2024.107588.

Dias, S. M. (2016). Waste pickers and cities. *Environment and Urbanization*, 28(2), 375–390. https://doi.org/10.1177/0956247816657302.

Diggle, A. (2022). *Exploring the capacity of the extended producer responsibility principle to incentivize the collection and recycling of plastic food packaging waste* (PhD diss.). https://dalspace.library.dal.ca/items/0ab221c9-7893-4479-8d35-7bb024ba3104

Environment and Climate Change Canada. (2023). *Canadian environmental sustainability indicators: Solid waste diversion and disposal*. www.canada.ca/en/environment-climate-change/services/environmental-indicators/solid-waste-diversion-disposal.html.

Ernstson, H., Lawhon, M., Makina, A., Millington, N., Stokes, K., & Swyngedouw, E. (2021). Turning livelihood to rubbish? The politics of value and valuation in South Africa's urban waste sector. In M. Keith & A. A. de S. Santos (Eds.), *African cities and collaborative futures: Urban platforms and metropolitan logistics* (pp. 96–120). Manchester University Press.

Gutberlet, J. (2023). Global plastic pollution and informal waste pickers. *Cambridge Prisms: Plastics*, 1(9), 1–16. https://doi.org/10.1017/plc.2023.10.

Gutberlet, J., Carenzo, S., Kain, J. H., & Azevedo, A. M. M. de. (2017). Waste picker organizations and their contribution to the circular economy: Two case studies from a Global South perspective. *Resources*, 6(4), 52. https://doi.org/10.3390/resources6040052.

Housing Needs Assessment. (2021). *Income categories and affordable shelter costs*. https://hart.ubc.ca/housing-needs-assessment-tool.

International Solid Waste Association. (2024). *Global waste management outlook 2024: Beyond an age of waste, turning rubbish into a resource*. https://wedocs.unep.org/bitstream/handle/20.500.11822/44939/global_waste_management_outlook_2024.pdf?sequence=3.

Kain, J. H., Zapata, P., Mantovani Martiniano de Azevedo, A., Carenzo, S., Charles, G., Gutberlet, J., … & Zapata Campos, M. J. (2022). Characteristics, challenges and innovations of waste picker organizations: A comparative perspective between Latin American and East African countries. *Plos One*, 17(7), e0265889. https://doi.org/10.1371/journal.pone.0265889.

Kalina, M. (2020). Waste management in a more unequal world: Centring inequality in our waste and climate change discourse. *Local Environment*, 25(8), 612–618.

Lee, M., Li, B., Maxwell, S., & Shulman, T. (2021). *A zero waste agenda for BC*. CCPA-BC and Zero Waste BC. www.zerowastebc.ca/wp-content/uploads/2021/10/ccpa-bc_Zero-Waste_2021_full.pdf.

Mesquita, J. L. C., Gutberlet, J., Araujo, K. P. de, Cruvinel, V. R. N., & Duarte, F. H. (2023). Greenhouse gas emission reduction based on social recycling: A case study with waste picker cooperatives in Brasília, Brazil. *Sustainability*, 15(12), 9185. https://doi.org/10.3390/su15129185.

Metro Vancouver. (2010). *Integrated solid waste and resource management: A solid waste management plan for the Greater Vancouver Regional District and member municipalities*. https://metrovancouver.org/services/solid-waste/Documents/integrated-solid-waste-resource-management-plan.pdf.

Metro Vancouver. (2022). *2021 Biennial Report. Integrated Solid Waste and Resource Management Plan.* January 11, 2022. Revised on November 2, 2023. Available at: https://metrovancouver.org/services/solid-waste/Documents/iswrmp-biennial-report-2021.pdf. Access on 19 Jun. 2024.

Phelan, A. A., Meissner, K., Humphrey, J., & Ross, H. (2022). Plastic pollution and packaging: Corporate commitments and actions from the food and beverage sector. *Journal of Cleaner Production*, 331, 129827.

Porter, M. E. (2015). Marginal recycling: Place and informal recycling in St. John's, Newfoundland. *Local Environment*, 20(2), 149–164. https://doi.org/10.1080/13549839.2013.836162.

Recycle BC. (n.d.). *About Recycle BC*. https://recyclebc.ca/about.

Tremblay, C. (2007). *Binners in Vancouver: A socio-economic study on binners and their traplines in downtown eastside* (Master dissertation, University of Victoria). http://hdl.handle.net/1828/997.

United We Can. (n.d.). *We produce results for our customers and community*. www.unitedwecan.ca/.

Zisopoulos, F. K., Steuer, B., Abussafy, R., Toboso-Chavero, S., Liu, Z., Tong, X., & Schraven, D. (2023). Informal recyclers as stakeholders in a circular economy. *Journal of Cleaner Production*, 415, 137894. https://doi.org/10.1016/j.jclepro.2023.137894.

9 Recycling right

A multi-stakeholder perspective from Ontario

Scott Cameron Lougheed and Virginia Maclaren

Introduction

Recycling has become a well-accepted part of most Canadians' daily lives and a major way that many feel they can reduce their environmental impact. Since curbside recycling was introduced to Canadians in 1981 with a pilot project in Kitchener, Ontario (Stewardship Ontario, 2013), residents have become diligent sorters of their discards while the array of products deemed "recyclable" grows rapidly. A dramatic shift in global recycling markets in 2018 drew attention to the issue of contamination of the municipal recycling stream. Contamination refers to items placed in a recycling bin that are not accepted by the local recycling system. In 2018, demands by materials processors in China and elsewhere for a less contaminated stream of recyclable materials upended the financial viability of recycling programs across Canada (Bula, 2018) and placed pressure on households to *recycle right*.

At about the same time, the model of recycling programs run by local governments in the province of Ontario was being rethought. Recycling of packaging and printed paper suffered from a growing number of problems in the province: the rate of recycling recovery had stagnated, and both contamination of the recycling stream and the cost of running the residential recycling program (known as the Blue Box program) had been rising steadily (Resource Productivity and Recovery Authority, 2023). In 2016, Ontario introduced the Waste Free Ontario Act (Province of Ontario, 2016), laying the groundwork for suppliers (collectively referred to as producers) of Blue Box material to the Ontario market to assume full, individual[1] responsibility for funding and operating the province's municipal recycling programs. This represented a shift from a model where they had only been responsible for about half of the costs of the program and municipalities bore the other half. Municipalities were also responsible for all operational decisions under the old model, including which recyclables to collect, how to educate residents about sorting, and whether to collect recyclables in a single stream or multiple streams.

Under the new model, known as full extended producer responsibility (EPR), producers will be required to collect a set list of packaging types across the province, will have responsibility for promotion and education, and will decide what

DOI: 10.4324/9781003534129-13

system to use for collecting recyclables (RPRA, 2024). Full transition to the new EPR model will be complete by 2025. Ontario will then become the second province, after British Columbia (Canada's Ecofiscal Commission, 2018), to have full EPR for residential recycling in Canada. All communities in the province will be eligible for collection services from producers, even those that do not currently have a recycling program. One exemption is for municipalities and First Nation communities in the province's Far North (RPRA, 2024). The Far North comprises 42% of Ontario's land mass but has a population of only 24,000 (Government of Ontario, 2023). Transport costs are a significant barrier to establishing recycling programs in remote, northern communities (ECCC, 2017).

In this chapter, we draw on interviews with packaging and recycling stakeholders to examine the reasons for contamination of the Ontario recycling stream and explore what the ongoing shift in governance of the province's recycling system might mean for the role of the diligently sorting resident. We employ the concept of consumption work (Wheeler & Glucksmann, 2015a) in highlighting the extent to which the recycling industry relies on the waste work of consumers to sort recyclables and four challenges that consumers face in *recycling right*. We pay most attention to packaging, and less to printed paper such as newspapers and magazines, because of the greater difficulty of sorting packaging of different material types compared to the printed paper stream.

Contamination of Ontario's Blue Box

As of 2023, Ontario's Blue Box program was providing 97% of Ontario's residents with access to recycling (RPRA, 2023). However, it had a contamination problem. The three main sources of contamination in the Blue Box were:

a recyclable but difficult to sort or market materials (e.g., black plastics, plastic/laminated pouches);
b degraded or contaminated recyclable materials (e.g., wet fiber, food left in a container); and
c prohibited materials (e.g., batteries, medical waste) (EcoCompass Consulting, 2020).

Despite efforts by municipalities to reduce contamination with education campaigns, Blue Box inspections, refusal to collect contaminated Blue Boxes and even fines (e.g., City of Toronto, 2018; HDR Corporation 2020), contamination remained stubbornly high. The highest contamination rates were found in Ontario's largest municipality, Toronto, and the largest regional municipality, Peel, with rates of 30% and 38%, by weight, respectively (Rider, 2023; Region of Peel, 2023).

Contamination of recycling streams makes it difficult to market sorted recyclables and capture revenues that help offset program costs. In 2018, contamination became a critical issue for municipalities everywhere when China implemented its National Sword Policy. The policy banned the import of mixed paper and plastics and imposed a dramatic 0.5% contamination limit on materials that it would accept (City of Toronto, 2018; Gregson & Crang, 2018). To that point, China had been one of the world's largest buyers of recyclable material. The China policy caused the prices offered on global markets for recovered recyclables to fall precipitously. In Ontario, it resulted in large revenue losses for municipal recycling programs, where revenues declined by over 50% between 2017 and 2020 (RPRA, 2023). Prices started to recover in 2021 as municipalities found new markets that tolerated higher contamination levels (RPRA, 2023), but contamination has remained a concern.

Consumption work

Wheeler and Glucksmann (2015a, p. 37) define consumption work as "all work necessary for the purchase, use, reuse and disposal of consumption goods and services". In undertaking the disposal task, consumers provide unpaid sorting, storage, and distribution services for downstream recycling industries and beyond (Anderson & Brodin, 2005; Jalil et al., 2016). Conceptualized in this way, consumers are best understood as a "pivot point node" between forward and reverse supply chains, rather than the end of the supply chain (Jalil et al., 2016, p. 246), creating new economic opportunities downstream by providing feedstock for recyclers and manufacturers, which in turn creates jobs and profits (Wheeler & Glucksmann, 2015b). Downstream stakeholders in the reverse supply chain depend on the labor of consumers to provide feedstock that can be profitably sold on the recycling market. However, unlike conventional suppliers and distributors, consumers typically pay for the service they provide, such as through collection fees and free time, rather than getting paid (Anderson & Brodin, 2005).

As Wheeler and Glucksmann (2015a) illustrate in case studies of Sweden and the UK, the configuration of the reverse supply chain can impact the consumption work of household recycling by raising uncertainty about the recyclability of certain types of packaging. They give the example of how the technology of a local Material Recycling Facility (MRF) determines whether a juice carton is or is not recyclable. MRFs employ hand-sorting and a variety of technologies to sort materials (e.g., optical, electromagnetic and eddy current sorters) (Gundupalli et al., 2017) but not all MRFs have the sophisticated technology needed to sort multi-material packaging like juice cartons.

Wheeler and Glucksmann (2015a) also comment on how other packaging characteristics affect consumption work. They give examples of the often time-consuming work required to prepare packages for recycling, including rinsing, removing sticky residue, flattening boxes, and pulling packages apart to separate materials. They also point to labels on packaging that claim recyclability but the community's recycling system cannot recycle that type of packaging. Plastics can be particularly frustrating to sort because so many different types of plastic are used in packaging, some of which may be recyclable and some not. A US study found that most consumers are uncertain about the meaning of the number found within the chasing arrows symbol on plastic packaging, even though it is supposed to help identify recyclable plastics (Latkin et al., 2022). This type of confusion is present among Canadian consumers as well and has spurred the federal government to develop legislation that will regulate labels. It aims to introduce standardized labels and allow producers to label plastic packaging as recyclable only if the package is accepted in a recycling collection system accessible to at least 80% of the population in a province or territory (Government of Canada, 2023).

Methods

To collect data for this study, we conducted semi-structured interviews with 24 stakeholders: waste management and recycling industries (4), municipal government and municipal government associations (7), nongovernment organizations (2), consultants (2), and producers and producer associations (9). Sampling was a combination of opportunistic and snowball sampling, with 13 initial participants derived from a review of 37 stakeholder organizations mentioned in reference to Ontario waste policy in public documents and the media. The interview guide focused on reasons for contamination of materials in the recycling program, how to remediate the contamination issue, and views on the impact of the proposed transition to full EPR. Interviews were approximately one hour long and were conducted in accordance with University of Toronto Ethics Protocol #36320. Interview participants received an information letter about the study, signed a consent form before the interview and agreed to be identified by sector in the results, but not by name. Transcripts were analyzed in QSR International's NVivo 12. A first round of open coding was used to identify general themes such as contamination, recycling markets, packaging issues, EPR, and the roles of consumers, municipalities and producers in the reverse supply chain, while secondary coding focused on clarifying initial themes.

Results and discussion

Many study participants mentioned how China's National Sword Policy changed the landscape for recycling in Ontario because of the new contamination standards.

> What is considered contamination today was perfectly acceptable material even 5 or 6 years ago because … people were willing to take this material. And now those markets are saying we don't want your trash, and that's everywhere, not just China. So, our markets are severely limited, and our definition of contamination is changing in that sense.
>
> (Municipal Government C)

Participants identified four main reasons for high contamination levels in Ontario: geographic variability of accepted materials, type of collection system, nonrecyclable packaging and recycling complexity. We outline these factors in the sections that follow and illustrate how they affect consumption work.

Geographic variability of accepted materials

Participants unanimously felt that geographic variability for materials accepted in the Blue Box across the province added complexity to Blue Box utilization and contributed to contamination. Currently, each municipality in Ontario can choose what materials it will collect from households in its Blue Box, meaning that households receive different instructions about what is recyclable in different municipalities. Recycling instructions can even differ within a municipality because commercial spaces such as malls, and office buildings do not have municipal collection. Instead, they contract with the private sector for garbage and recycling collection and those contractors may not collect the same materials as the municipality. One respondent summarized the problem as follows:

> The research we've done with consumers indicates that consumers are very motivated to recycle. But it's very confusing for consumers to recycle, particularly in Ontario, because of the very fact that if you have a cottage or if you've travelled outside of Toronto and stayed at somebody else's house in another municipality, you'll know that municipalities collect different things. So, it's not surprising that consumers would be quite confused about what can and cannot be recycled.
>
> (Producer A)

Because Ontario residents often live, work, and play in different municipalities, remembering what is applicable in each can be difficult. Instructions for recycling take-out coffee cups and lids are an example mentioned in several of the interviews. Table 9.1 shows the instructions for recycling coffee cups and lids in five contiguous municipalities in Southern Ontario. Only two of the municipalities located next to each other—Burlington and Oakville—have the same instructions. The reason is that only two municipalities have the level of composting technology needed to compost coffee cups from Green Bins and only one has the recycling technology at MRFs required to sort coffee cups from the Blue Box.

Table 9.1 Managing take-out paper coffee cups in contiguous Ontario municipalities

Municipality	*Population 2021*	*Paper coffee cup*	*Plastic coffee cup lid*
Hamilton	569,400	Blue Box	Blue Box
Burlington	187,000	Green Bin (Organics)	Blue Box
Oakville	214,000	Green Bin (Organics)	Blue Box
Mississauga	718,000	Garbage	Garbage
Toronto	2,794,000	Garbage	Blue Box (except black plastic lids, which go in Garbage)

Source: This was developed using information from Ontario municipal recycling websites and Statistics Canada (2022).

Type of recycling collection system

The two types of recycling collection systems used in Ontario are single stream and dual stream. A single-stream system allows residents to put all products accepted by the recycling system into a single curbside bin. Dual stream systems require households to sort paper fibers (paper and paper packaging) into one bin and plastics, metals and glass in another. Several participants commented on how the single-stream system lowers collection costs for municipalities because only one bin rather than two bins is being collected. They also mentioned how the system requires less sorting by households and less storage space inside and outside the home. However, they emphasized that a drawback of the single-stream systems is that it usually has higher contamination rates because it places a higher burden on MRF sorting technologies and can contaminate paper fibers with liquids and broken glass slivers. From a consumption work perspective, a single-stream system places more of the burden for sorting materials on the paid labor and infrastructure of MRFs while a dual stream system relies more heavily on the unpaid labor of individual residents (Wheeler & Glucksmann, 2015b).

Two research studies in Ontario confirm the lower collection costs and higher contamination and processing costs of single-stream systems (HDR Corporation, 2013; Lakhan, 2017). Because of the higher contamination rates, municipalities with single-stream systems had greater difficulty marketing their recyclables after the world recycling market disruption in 2018, as noted by this participant.

> When China dramatically changed the rules of the game … it was a significant issue because dual stream operations had fewer problems moving material and were more successful in being able to market material than single stream facilities … Many of the dual stream facilities maybe had to slow down the line a little more or do a little bit of extra quality improvements but it wasn't as dramatic as single stream … a lot of the single stream facilities were having trouble marketing their material or they had to run it through the facility once or twice or just throw out more material.
>
> (Recycling Industry B)

Strictly speaking, type of collection system does not affect a resident's ability to recycle right a great deal. It is easier to put all packaging and paper into one bin rather than two, but the real contamination problems of single-stream systems happen during transport and processing when there is a higher incidence of broken glass and cross-contamination occurs between packaging and paper (Berardocco et al., 2022).

Nonrecyclable packaging

Consumer confusion about recycling is further compounded by the introduction by producers of nonrecyclable packaging that may resemble commonly recyclable packaging and have symbols or text claiming that they are recyclable. These claims about recyclability may not be true in a municipality where appropriate infrastructure to handle that novel or complex package is lacking. Some participants—mostly from government, NGO, and recycling sectors—referred to the misleading recycling claims from a few brands as a deliberate effort to greenwash their products.

> [Consumers] get confused because there are so many packages on the store shelf that appear to be recyclable, or that the retailers say are recyclable but aren't really accepted in the programs ... And that leads to some of them throwing stuff in the garbage instead of the Blue Box because they're not sure what to do, and some throwing stuff in the Blue Bin that should be in the garbage.
>
> (Municipal Government D)

One example of confusing packaging was cited regularly by participants: single-serving coffee pods. They are not recyclable in most municipalities in Ontario and epitomize the problem of misleading claims.

> [A coffee pod brand] just ran a big national campaign saying "our coffee pods are recyclable," even though the majority of municipalities don't actually accept them. To me that's false advertising and that shouldn't happen. But that's the type of thing that, they're getting educated in a particular way. It is recyclable I guess, if there's a location available, but there's not.
>
> (Consultant A)

Proposed federal government labeling requirements for plastics packaging should help address this problem. Although the labels will only be required for plastics, producers will have the option of using them on other types of packaging as well. The labels are a work in progress, not yet approved by the federal government at time of writing but may look similar to the examples shown in Figure 9.1. The QR code would link to a website that identifies which provinces and territories

Figure 9.1 Examples of potential recyclability labels for single and multi-material plastic packaging.

Source: This was adopted from the Government of Canada (2023).

accept the item for recycling, provides an explanation of why the item is recyclable or not, and explains how to prepare an item for recycling, such as by separating components (Government of Canada, 2023).

Although better labels will help reduce consumer confusion, another issue with the single-serve coffee pod is that it normally requires considerable effort to disassemble and prepare them for recycling, including peeling off the foil top, which may or may not be recyclable, removing the spent coffee grounds, placing them in the garbage or compost (if available), and placing the plastic cup portion in the recycling bin.

> [Single-serve coffee pod manufacturers] have given the impression that the consumer can recycle it, and in a perfect world, if you follow steps 1, 2, 3, 4, and 5, you absolutely could recycle the lion's share of the material component parts of that pod. The reality is that very few people are going to follow those steps. They're going to toss it because you bought it for convenience, you're probably going to toss it for convenience.
>
> (NGO A)

Even when consumers follow these steps, participants, particularly from local government, noted that the increased standards for contamination and dominance of single-stream recycling could still lead to small amounts of coffee or coffee grounds staining other material in the single-stream Blue Box, such as paper, rendering that material nonrecyclable. This latter point highlights how infrastructure and system design (e.g., single-stream recycling), not just consumption work, play a significant role in contamination rates.

Responding to concerns about nonrecyclable packaging, four producers pointed out that many new packaging types are chosen for other benefits, such as extending shelf life of the product and lowering transportation costs and emissions.

> We [sell our product] in a multi-laminate pouch … and there's hardly any material there to transport a kilogram of [the product]. Even though it's not recyclable, and the other ones are highly recyclable, it beats out other packaging five to ten times on various environmental factors. [Some say] "I don't care, I just want to recycle." And I say, "yeah, you want to recycle at any cost to the environment, any cost to the economy." So, I think there's an inherent lack of understanding of science.
>
> (Producer E)

This quote highlights the producer view that switching from a lighter nonrecyclable package to a heavier recyclable package would be more environmentally harmful, and that science (e.g., a lifecycle assessment of the package) supports the producer position. The same argument about environmental benefits was offered by another producer when comparing nonrecyclable packaging to no packaging at all. The example given was nonrecyclable plastic wrap around a cucumber, which extends the product's shelf life from 4 to 14 days and reduces food waste, another important environmental concern (Producer A). In contrast to the perspective of producers, three participants (from municipal government and consulting) argued that choosing a packaging material that has these upstream environmental benefits should not "give brand owners a pass" (Municipal Government E) on the lack of recyclability.

Recycling complexity

Recycling complexity refers to the complex decisions consumers often face when trying to determine whether a package is recyclable or nonrecyclable. Many participants in the study commented on the limits of individual agency and capacity for making those decisions. For some participants this was a sign of deficiencies or failures of individuals.

> I must admit, sometimes I thought "I'm not sure if this is recyclable or not" but in the moment, you don't go to your computer and check … it's a flick of the wrist. You make a decision based on the information you have or how lazy you are, right? You flick your wrist either way. Do you rip off the little bit of boxboard, do you rip the plastic facing off? … Because really you should, because paper and plastic are different. Some people just throw the same thing into the garbage or into the recycling.
>
> (Producer D)

This quote suggests that laziness and lack of information at the moment of discard may lead individuals to make a spontaneous and incorrect decision, rather than taking the time to look at local government resources to determine where the material belongs. It places blame for contamination on consumers for not properly sorting, rather than on producers for choosing problematic packaging. Another view of how the consumer deals with recycling complexity comes from a local government participant:

> There are those people … that practice "wishcycling." They don't want to respect what the local council has decided to expend funds on to recycle at the curb and they want to try and expand that [because] they see a symbol on the bottom … So, they look at that and say "This is recyclable" and so they put it in the Blue Box … And you know they wish even if it's not, they wish that it was, and they'll put it in anyway. It just increases cost.
>
> (Municipal Government A)

Rather than being lazy, this participant views individuals as possessing agency to do what they are instructed by the package itself, or actively defying instructions from municipalities in the hopes of inciting a desired change.

Several participants declared that if the majority of consumers are consistently having difficulty properly handling a specific type of packaging, perhaps it is the packaging that requires modification.

> [Improper recycling] is a pattern of behaviour that we need to understand—and not necessarily modify! It's the packaging that needs to be modified.
>
> (Recycling Industry A)

> Is the product wrong? Is the package wrong? That's where you want innovation to happen. If it's actually that difficult to handle, then maybe the incentive is let's get out of doing it that way. Let's package it in a different way, let's build it out of a different material. How do we create those right signals … you need a nudge in the right direction and we don't have that right now.
>
> (Consultant A)

These quotes reflect a view shared by a number of participants from outside the packaging sector that producers should design packaging that better suits local infrastructure and local resident capacities for sorting.

Shifting to full extended producer responsibility

To what extent will shifting to full EPR affect consumption work and alleviate contamination? Under the new system, there will be harmonization of materials collected in the Blue Box across the province. There was near consensus among study participants that harmonization will help consumers recycle right.

> I think that if you have more consistency between what is considered acceptable or not, if I were to move from Simcoe county, to Toronto, to Halton region, I wouldn't have to worry about understanding what I can toss in my Blue Bin or not. I think it would be a little bit easier for people to not have to go through the extra effort of checking all the time. Right? It would be more of an engrained behaviour. So, I think that would make it just easier across the board.
>
> (Producer G)

Harmonization may require upgrading of MRFs in many parts of the province to handle new types of packaging and meet what are expected to be stringent contamination standards set by producers. Producers want low contamination rates because it will allow them to market and derive revenue from as much recovered material as possible rather than sending contaminated loads to landfill.

British Columbia offers an example of stringent contamination standards under full EPR. If a municipality in the province has a recycling collection contract with producers, contamination of recyclables that the municipality collects can be no higher than 3% or else a municipality is subject to fines (Regional District of Central Okanagan, 2023). Although this contamination rate may seem incredibly low from an Ontario perspective, the EPR system in BC has several advantages over Ontario's current system: (1) it is a province-wide harmonized system for recycling packaging types; (2) education programs target the same types of packaging for all communities, with some minor exceptions; (3) it has a deposit-return system for virtually all beverage containers, keeping them out of the Blue Box and simplifying sorting; and (4) some packaging that is difficult to sort at MRFs (glass, plastic film and foam packaging) has been removed from the Blue Box to depot collection or, in some cases, to a separate collection system at curbside (Continuous Improvement Fund, 2020).

Returning to Ontario, another benefit of harmonization is that it will allow for the delivery of consistent province-wide education programs and this was seen as a major plus of the new model, as evident in the comment below.

> Today, when you think about how media crosses municipal boundaries, a resident might be seeing two or more instructions about what to do with certain types of packaging, and those instructions could be in conflict. So, once we get to a standardized list of materials targeted across the province, we'll be able to standardize communication out to the residents of what to do. That will reduce contamination, it will reduce wishful recycling.
>
> (Producer C)

One dissenting voice about the value of harmonization felt that it could result in push-back among residents in some municipalities if any packaging types are removed from their Blue Boxes and shifted to another form of collection, such as a depot system. In British Columbia, it took about a year in some municipalities for residents to change their sorting behavior when plastic film and glass were no longer accepted in the Blue Box (Continuous Improvement Fund, 2020). If, under full EPR in Ontario, some materials were to be pulled out of the Blue Box and directed to other collection streams, consumption work would change since recycling the materials would require separate storage in the home and transporting them to depots.

Participants were not optimistic that full EPR would encourage producers to move away from complex and nonrecyclable packaging. Theoretically, it will be in the producers' interests to make it as easy as possible for consumers to prepare and correctly sort packaging because producers will be responsible for recovering a legally determined percentage of the packaging they sell. However, as one producer noted:

> I'm not sure [full EPR] will actually have a significant impact on packaging design because the first role of packaging is to protect the product ... And environment and end-of-life is one of several considerations. I don't necessarily think [packaging designers] will be completely changing their packaging design, but they might be sharing more information [with MRF operators and recyclers] ... to make it more compatible with the recycling system.
>
> (Producer H)

Three producers mentioned that an important factor influencing design is competition for innovative packaging that will help producers sell their products. Another producer remarked that in larger organizations, those who manage stewardship fees and those who design packaging may not be in close communication. Producers that operate internationally commented that even in jurisdictions with relatively advanced systems of EPR such as the EU, the incentives to rethink package design were lacking. Joltreau (2022) supports this contention, finding very little evidence in 25 European countries that producers subject to EPR fees have modified their packages to reduce packaging or make them more recyclable. With the exception of two producers, who held neutral views, producers in the current study were eager to take on responsibility for the recycling system, in large part because of their conviction that they can operate it more efficiently than local governments. Producers were less convinced that EPR would result in the reduction of complex and nonrecyclable packaging because of the multiple factors that affect packaging design.

Conclusion

Contamination of the residential recycling stream is a significant problem in Ontario with multiple causes. As identified by stakeholder interviews, they include geographic variability of accepted materials, nonrecyclable packaging, recycling complexity and type of collection system. All of these, the first three in particular, affect consumption work and the ability of Ontario residents to *recycle right*.

Harmonization of packaging accepted for collection under full EPR is expected to address some of these issues. It will help alleviate contamination problems related to geographic variability and recycling complexity. The same items will be recyclable in all municipalities, reducing uncertainty about sorting. Second, following from the first point, producers can use the same education program province-wide rather than having to tailor it to specific recycling programs in specific municipalities. This should help improve awareness about the recyclability of different types of packaging and reduce recycling complexity to some extent. Harmonization will still leave some confusion because Ontario's full EPR is for residential recycling only, and the types of packaging accepted for recycling in offices, retail and institutions may be different.

Full EPR is not expected to result in wholesale redesign of problematic packaging so that it is easier for consumers to recycle, nor in the cessation of new nonrecyclable packaging arriving on the market. Full EPR may increase consumption work by transferring some materials from curbside Blue Box collection to depot collection. In sum, we find that while there is growing recognition among stakeholders that improving recycling performance requires action from producers rather than just households, for most, particularly those in the packaging sector, households remain the primary point of leverage for recycling right.

Note

1 "Individual" here refers to an individual company as opposed to collectives of companies sharing responsibility.

References

Anderson, H., & Brodin, M. H. (2005). The consumer's changing role: The case of recycling. *Management of Environmental Quality: An International Journal* 16(1), 77–86. https://doi.org/10.1108/14777830510574362.

Berardocco, C., Delawter, H., Putzu, T., Wolfe, L. C., & Zhang, H. (2022). Life cycle sustainability assessment of single stream and multi-stream waste recycling systems. *Sustainability*, 14(24), 16747. https://doi.org/10.3390/su142416747.

Bula, F. (2018). China's tough new recycling standards leaving Canadian municipalities in a bind. *Globe and Mail*, January 18.

Canada's Ecofiscal Commission. (2018). *Cutting the waste*. McGill University.

City of Toronto. (2018). *State of Toronto's Blue Bin Recycling Program*. www.toronto.ca/legdocs/mmis/2018/pw/bgrd/backgroundfile-113576.pdf.

Continuous Improvement Fund. (2020). *Part I. Learnings from British Columbia: Advice for transition*. https://thecif.ca/learnings-from-british-columbia-advice-for-transition.

EcoCompass Consulting. (2020). *Inbound contamination definition & impact analysis*. Continuous Improvement Fund Project # 1090.

Environment and Climate Change Canada. (2017). *Solid waste management for northern and remote communities*. Environment and Climate Change Canada.

Government of Canada. (2023). *Recycled content and labelling rules for plastics: Regulatory framework paper*. www.canada.ca/en/environment-climate-change/services/canadian-environmental-protection-act-registry/recycled-content-labelling-rules-plastics.html.

Government of Ontario. (2023). *Far North of Ontario*. www.ontario.ca/page/far-north-ontario.

Gregson, N. & Crang, M. (2018). Made in China and the new world of secondary resource recovery. *Environment and Planning A: Economy and Space*, 51(4), 1031–1040. https://doi.org/10.1177/0308518X18791175.

Gundupalli, S. P., Hait, S., & Thakur, A. (2017). A review on automated sorting of source-separated municipal solid waste for recycling. *Waste Management*, 60, 56–74. http://dx.doi.org/10.1016/j.wasman.2016.09.015.

HDR Corporation. (2013). *An assessment of single and dual stream recycling including current program performance in large Ontario municipalities*. Waste Diversion Ontario, Continuous Improvement Fund Office.

HDR Corporation. (2020). *Comparative scan of municipal strategies, practices and initiatives*. City of Ottawa, Solid Waste Master Plan, Technical Memorandum #3.

Jalil, E. E., Grant, D. B., Nicholson, J. D., & Deutz, P. (2016). Reverse logistics in household recycling and waste systems: A symbiosis perspective. *Supply Chain Management*, 21(2), 245–258 https://doi.org/10.1108/SCM-02-2015-0056.

Joltreau, E. (2022). Extended producer responsibility, packaging waste reduction and eco-design. *Environmental and Resource Economics*, 83(3), 527–578. https://doi.org/10.1007/s10640-022-00696-9.

Lakhan, C. (2017). A comparison of single and multi-stream recycling systems in Ontario, Canada. *Resources*, 4(2), 384–397. https://doi.org/10.3390/resources4020384.

Latkin, C. A., Dayton, L., Yi, G., & Balaban, A. (2022). The (mis) understanding of the symbol associated with recycling on plastic Containers in the US: A brief report. *Sustainability*, 14(15), 9636. https://doi.org/10.3390/su14159636.

Province of Ontario. (2016). *Waste Free Ontario Act, S.O. 2016. 12—Bill 151*. www.ontario.ca/laws/statute/s16012.

Region of Peel. (2023). *Update on Peel's current and future curbside and multi-residential enforcement programs.* Report submitted to Region of Peel Waste Management Strategic Advisory Committee.

Regional District of Central Okanagan. (2023). *Recycling contamination = Financial penalties.* Regional District of Central Okanagan News and Notices. www.rdco.com/en/news/recycling-contamination-financial-penalties.aspx.

Rider, D. (2023). *City staff think Toronto should get out of the recycling business. why? That's a secret.* Toronto Star, October 2.

RPRA. (2023). Datacall trends analysis of Ontario's residential recycling programs from 2002 to 2021. Resource Productivity and Recovery Authority. https://rpra.ca/wp-content/uploads/Datacall-Trends-Report.pdf.

RPRA. (2024). *Blue Box program operated under Ontario's new producer responsibility framework.* Resource Productivity and Recovery Authority. https://rpra.ca/programs/blue-box/regulation.

Statistics Canada. (2022). *Population and dwelling counts: Canada, provinces and territories, census divisions and census subdivisions (municipalities).* www150.statcan.gc.ca/t1/tbl1/en/tv.action?pid=9810000203&geocode=A000235.

Stewardship Ontario. (2013). *The story of Ontario's Blue Box.* Stewardship Ontario.

Wheeler, K., & Glucksmann, M. (2015a). *Household recycling and consumption work.* Palgrave Macmillan. https://doi.org/10.1057/9781137440440.

Wheeler, K., & Glucksmann, M. (2015b). It's kind of saving them a job isn't it? The consumption work of household recycling. *Sociological Review*, 63(3), 551–569. https://doi.org/10.1111/1467-954X.12199.

10 Recycling behaviors and knowledge in Western Newfoundland

Explanations and recommendations

Hadiya Bamragha and Garrett Richards

Introduction

Over the years, recycling has become a way of life for many people in Canada. Contamination in recycling bins continues to be a problem, however, and it is not yet clear what institutional strategies are most effective at improving accurate sorting by residents. In Western Newfoundland, the scope of our study, waste management is led by an organization called Western Regional Waste Management (WRWM), which has overseen a mandatory recycling program for single-unit households since July 1, 2018 (WRWM, 2019). Several measures have been taken by the region to ensure residents are recycling properly, such as fining violators and auditing samples of collected bags (WRWM, 2019).

In February 2019, an audit of clear and blue bags from 100 sample households found that about 7% of the material in clear bags did not belong there (e.g., recyclable paper, hazardous materials) and about 15% of the material in blue bags did not belong there (e.g., ceramics, glass; MMSB, 2019). Contaminated blue bags may be landfilled. Overall, the capture rate was estimated to be 78% for 2018; out of approximately 4,100 tons of recyclable material generated, roughly 3,200 tons were recycled and 900 tons were landfilled (MMSB, 2019). The 2019 audit results estimated the region's total diversion rate to be 19%, predicting that it could increase to 24% if residents recycled correctly, without contamination (WRWM, 2019).

Case study description

The Western Newfoundland region is comprised of 73,500 individuals and 40,544 households (Statistics Canada, 2023). A mandatory recycling program offers weekly collection service to most of the region's single-unit households (MMSB, 2019). The recycling process begins at curbside where residents place garbage (clear bags) and recycling (blue bags) for collection. One of the hauler's responsibilities is to check the waste before collection; if waste is improperly sorted (e.g., there is garbage in a blue bag), the hauler can implement a warning or a fine (Fracassi, 2022). Residents can also drop off garbage at transfer stations and special items (e.g., household hazardous waste, tires) at designated sites (WRWM, n.d.). Finally, refundables, while accepted in the blue bags, can also be

DOI: 10.4324/9781003534129-14

dropped off at green depots and some licensed private businesses (MMSB, 2019). The hauler delivers clear and blue bags from the curb to any of the six waste transfer stations in the region, which weigh the bags to determine fees charged to haulers and their clients, such as municipalities (Fracassi, 2022). Finally, bags are transported to their final provincial destinations like landfills and recycling depots (Fracassi, 2022).

Purpose statement and research questions

Contamination continues to be a problem in Western Newfoundland. Thus, there is need to analyze recycling behaviors, perceptions, and knowledge among residents to identify stress points causing potential noncompliance with recycling instructions. More specifically, this study addresses the following research questions:

1 What explains the (lacking) compliance of residents with the blue-bag program in Western Newfoundland?
  a Is it a matter of not recycling enough? If so, what factors explain the amount that households recycle?
  b Is it a matter of not knowing how to recycle properly? If so, what factors explain the recycling knowledge of residents?
2 What can be done, in terms of education and management, to increase compliance with recycling guidelines?

Factors influencing recycling behavior

There are several pieces of literature offering potential factors for consideration in the case of Western Newfoundland. Investigations of household recycling behavior (e.g., Lee & Khan, 2020; Yang & Arhonditsis, 2022) have unveiled a plethora of influences, prompting broad categorization. For instance, Miafodzyeva and Brandt (2013) categorized explanatory variables into three general groups: technical-organizational elements, sociodemographic considerations, and sociopsychological factors.

The first category is technical-organizational factors, which include: convenience, access to curbside collection, and unit pricing. Examples of convenience are frequency of collection, distance and location of collection points, transportation and handling issues, easy access, and storage space (Miafyodzyeva & Brandt, 2013). Numerous studies indicate that making recycling more convenient can be a straightforward way to boost recycling behaviors (DiGiacomo et al., 2018; Munro, 2010). Access to curbside recycling programs is a primary motivator in encouraging recycling activities (Lakhan, 2015; Rosenthal & Linder, 2021). Studies on the role of monetary incentives like unit-based pricing and weight-based pricing on recycling behavior have shown mixed results (Dahlén & Lagerkvist, 2010; Lakhan, 2015).

Secondly, sociodemographic factors also play a role in predicting recycling behavior. Some studies find a correlation between greater age and recycling habits

(Anderson & Krettenauer, 2021; Yang & Arhonditsis, 2022), though others do not (Guiot et al., 2019). Many studies show that education is correlated with recycling (Hoffmann & Muttarak, 2022; Stojanovski et al., 2020). While higher income levels are sometimes associated with increased recycling activity (Yang & Arhonditsis, 2022), other studies do not support this observation (Ferrara & Missios, 2005). The influence of gender presents mixed findings as well (Anderson & Krettenauer, 2021; Kennedy & Kmec, 2018).

Finally, sociopsychological variables include moral norms, social norms, legal norms, and environmental concerns. Moral norms are crucial in shaping recycling behavior, with individuals tending to recycle more when they sense a personal obligation (Krettenauer & Lefebvre, 2021). Some studies observe dependence between social pressure and recycling habits (Gifford & Nilsson, 2014; Smith et al., 2021). Legal norms also influence recycling behaviors (Viscusi et al., 2014). Environmental concern may be a strong predictor of recycling intention, which can predict recycling compliance (Fan et al., 2019). Overall, this review demonstrates several potential variables for investigation in the case of Western Newfoundland. Also, it appears that no academic study has previously been conducted on recycling habits in the region. Our study intends to reveal how Western Newfoundland compares to other jurisdictions.

Methodology and research design

Given that possible leverage points can be identified from comparable studies in other places, this study employs a deductive (i.e., hypothesis testing) component, with public surveys as the method. However, to ensure we can fully appreciate the unique context of the region, the survey also incorporates an inductive lens to explore the issue broadly, outside of the hypotheses. WRWM was also a partner in the research, to provide background expertise (e.g., suggestions for hypotheses and survey questions) and increase the potential impact of the findings.

Research hypotheses

We set the following hypotheses, based on the above review and suggestions from the study's external partner (WRWM):

- Hypothesis 1a (corresponding to Research Question 1a)—amount of recycling will correlate positively with: age, women, level of education, household income (i.e., sociodemographic variables), recycling knowledge (i.e., technical-organizational variable), and other environmental behaviors (i.e., sociopsychological variable).
- Hypothesis 1b (corresponding to Research Question 1b)—recycling knowledge will correlate positively with: age, women, level of education, household income (i.e., sociodemographic variables), perceived availability of information about recycling in the community, perceived usefulness of information about recycling in community, perceived clarity of labels on recyclable products (i.e.,

technical-organizational variables), and wishful recycling (i.e., psychological variable).

Survey structure

Based on the research questions, factors identified in the literature, suggestions of the partner (WRWM), and general principles of survey design, the survey was developed (in Qualtrics) to consist of these sections: eligibility questions, questions on recycling behavior, questions on factors influencing recycling behavior, questions on recycling knowledge (including a quiz table checking whether the respondent knows how to properly dispose of different materials), management questions, open-ended qualitative questions, and demographic questions. Note that many questions asked the respondent to answer on behalf of their household.

Survey implementation

The target population was residents of Western Newfoundland aged 19 or over. We distributed the survey to 39 general community Facebook groups for municipalities in the region (e.g., “Corner Brook Classifieds” and “Stephenville Classifieds”). After obtaining permission from the respective group administrators, the survey was posted publicly in each group on various dates in October 2022, closing at the end of November 2022. Because the population of residents aged 19 or over in Western Newfoundland is 61,340 (Statistics Canada, 2023), the target sample size for the survey was 382 participants, based on a 95% confidence level and 5% margin of error (Qualtrics, 2024).

Data analysis plan

The first step in our data analysis process was data cleaning. After closing the survey with 1,242 respondents, we examined the responses for validity (e.g., removing from the sample respondents who indicated a postal code outside of Western Newfoundland). In doing so, we uncovered a very large number of spoof responses and underwent addition steps to purge them from the data (e.g., checking for combinations of internal inconsistency, unrealistic numbers for bags, multiple unusual responses in the quiz table, unintelligible qualitative responses, and suspicious patterns in consecutive responses). The final number of valid responses included in the analysis was 206. Because this number is less than the target sample size of 382, we should be cautious not to generalize the sample characteristics to the entire population.

The second step in our data analysis process was to consider representativeness. We directly compared demographic characteristics of the sample and the population, using census data for Western Newfoundland from Statistics Canada (2023). The comparison revealed some discrepancies, especially the underrepresentation of men, seniors, the Southwest region, and the town of Deer Lake in the sample. These biases may limit the generalizability of the findings to the population, but the extensive geographical coverage offers a foundation for identifying potential trends around recycling habits in the region.

With the data cleaned and the limitations understood, the third step in our data analysis process was to perform the core analysis necessary to answer our research questions. Using Excel, we generated descriptive summaries (e.g., averages, ranges, and histograms) for all the quantitative responses. Using SPSS, we conducted statistical analyses (e.g., multiple linear regressions) on the quantitative responses relevant to the hypotheses. Using Excel and Word, we conducted a qualitative analysis on the open-ended questions, categorizing the responses into themes. We did not decide on the themes in advance, instead allowing them to emerge from the data.

Results

This section reports the full details of relevant descriptive, statistical, and qualitative results for the study, presented in order of the research questions.

Recycling amount (Research question 1a)

A key question on the survey was a table of various factors that might influence how much a person recycles. For each factor, we asked respondents to indicate an answer on a five-point Likert scale from "strongly discourages my recycling" to "strongly encourages my recycling". Counting "strongly discourages" as double the weight of simply "discourages" (and same for the encouragement side of the scale), Table 10.1 is a descriptive summary of the ratio of encouragement to discouragement across the respondents for each factor, as well as the percentage of participants responding "neither" or "unsure" for context. Notable self-reported encouraging factors included "responsibility to future generations" (23.40 times as much encouragement as discouragement, or encouragement ratio), "importance of doing things properly" (22.00 encouragement ratio), "interest in limiting environmental impacts" (18.29 encouragement ratio), and "interest in supporting a green economy" (15.14 encouragement ratio). The most notable self-reported discouraging factor was "trust in effectiveness of recycling system" (0.96 encouragement ratio), the only factor to report more discouragement than encouragement from respondents.

Qualitative responses to the open-ended question "What encourages or discourages you to recycle?" provide some additional context. Out of 110 comments, 77 were negative (referring to discouragement), 20 were positive (referring to encouragement), and 13 were neutral or impossible to classify. Major themes around discouraging factors included: distrust that the system is actually recycling blue-bag materials effectively (24 comments), insufficient range of materials (e.g., glass) accepted (17 comments), inconvenience due to overly strict requirements for cleanliness and bag sorting (14 comments), lack of clear information on what is recyclable (12 comments), and limited or inconsistent access to curbside recycling and depots (nine comments). For example, one participant stated "when you hear rumors that the recycling bags are dumped or buried and

Table 10.1 Respondent self-reported influence of various factors on recycling participation

Potential influential factor	*Ratio of encouragement to discouragement across responses*	*% of participants indicating neither encouragement or discouragement (or unsure/no answer)*
Responsibility to future generations	23.40	22%
Importance of doing things properly	22.00	21%
Interest in limiting environmental impacts	18.29	14%
Interest in supporting a green economy	15.14	22%
Identification as an "environmentalist"	6.78	50%
Availability of recycling services	4.32	30%
Availability of recycling information	4.24	38%
Personal costs and benefits of recycling	4.00	48%
Enforcement of recycling requirement	3.50	49%
Usefulness of recycling information	3.20	38%
Clarity of labels on recyclables	2.54	37%
Overall convenience of recycling	2.31	24%
Pressure from social circle	1.67	82%
Trust in effectiveness of recycling system	0.96	33%

not actually recycled is discouraging" (distrust theme) and another said "I am on a septic system and having to clean my recycling is frustrating" (inconvenience theme). The only major theme around encouraging factors was responsibility to the environment or future generations (eight comments). These self-identified factors, especially those representing discouragement, are potential areas of intervention to increase participation in Western Newfoundland recycling programs, in particular improving trust in the system, which was a key focus of both the quantitative and qualitative responses.

However, the above questions focused only on self-reported factors. A statistical analysis was used to test Hypothesis 1a and look for relationships between potential causal factors and recycling amount. As a measure of recycling amount, we asked respondents to estimate how many clear (garbage) and blue (recycling) bags their household produces on an average week. We found that most common weekly number for both bag types is 1, but it is far more likely for households to produce multiple clear bags than multiple blue bags. Then, we calculated a rough recycling proportion for each household (i.e., number of blue bags divided by number of total bags); they ranged from 0% to 80%, with an average of 40%. This recycling proportion became the dependent variable for our statistical test (multiple linear regression), with numerous independent variables measured by other survey questions (i.e., age, gender, level of education, household income, recycling knowledge, other environmental behaviors). However, none of the relationships turned out to be significant, so the statistical output is not included in this piece. Age was the closest to being significant, with a p-value of 0.079 (still not meeting the significance threshold of 0.05); the others all had very high p-values over 0.25.

Hypothesis 1a was not supported. From the range of recycling proportions itself, though, we can still conclude that there is a wide variability across respondents, which highlights a potential need for targeted interventions to address the disparity in recycling behaviors.

Recycling knowledge (Research question 1b)

The most relevant survey question on recycling knowledge was a quiz table that asked respondents to indicate, for various sample items, whether it should be placed in the clear (garbage) bag, the blue (recycling) bag, or taken to a special collection site. We determined the correct answers using WRWM's publicly available recycling guide (see https://wrwm.ca/recycling-at-home/). Table 10.2 is a descriptive summary of the percentage of respondents answering correctly for each item and the distribution of correct answers across respondents. Respondents showed a good understanding of how to dispose of common items such as pop cans (100% knew to put them in the blue bag or take them to a special collection site), tires (97% knew to take them to a special collection site), and diapers (96% knew to put them in the clear bag). However, there was confusion around items like air freshener spray containers (only 28% knew to take them to a special collection site), wine bottles (only 45% knew to take them to a special collection site), and ceramic flowerpots (only 58% knew to put them in the clear bag). Items with low numbers of correct answers may merit more attention in educational materials. The overall scores of respondents ranged from 9/22 to 22/22, with most scores in the range of 18–21. The variation in scores among respondents is critical, as misunderstandings directly impact contamination rates during recycling, and even a single misunderstanding can lead to contamination.

For the statistical analysis (i.e., a multiple linear regression) of Hypothesis 1b, we used the quiz scores as the dependent variable, measuring revealed recycling knowledge. Various independent variables measured by other survey questions included: age, gender, education, household income, perceived availability of information about recycling in the community, perceived usefulness of information about recycling in community, clarity of labels on recyclable products, and wishful recycling (i.e., tendency to place items in the blue bag when unsure of where they should go). Table 10.3 depicts the results of the multiple linear regression.

The model's overall predictive power is low, only explaining a small percentage of the variability in revealed recycling knowledge across respondents, as evidenced by the low R-squared values. Still, the analysis does reveal a few significant influences (indicated by an asterisk and a p-value less than 0.05). The strongest and most significant factor was wishful recycling, with a B of –0.602 and a p-value of less than 0.001. These numbers mean that, for any participants who indicated that they would put an item they are unsure about in the blue bag, hoping that it is recyclable (instead of giving another answer to that question), we can predict that their revealed recycling knowledge score would decrease by 0.602 out of 22; the chance of this observed relationship being due to randomness is less than 0.1%, so it is significant. Gender was a significant predictor (B = –0.371; p-value = 0.011), with

Table 10.2 Results of recycling quiz survey question (measurement of recycling knowledge)

Item	*% of respondents answering correctly*	*# Correct answers (out of 22 Items)*	*% of respondents with that score*
Pop can	100%	22	2%
Plastic drink bottle	99%	21	11%
Cardboard box	98%	20	20%
Tire	97%	19	12%
Juice box	96%	18	17%
Diaper	96%	17	10%
Egg carton	94%	16	8%
Cell phone	94%	15	9%
Newspaper	91%	14	7%
Metal soup can	88%	13	3%
Bricks	86%	12	1%
Milk carton	85%	11	1%
Styrofoam packaging	84%	10	0%
Plastic yogurt container	83%	9	1%
Wax-coated take-out container	79%	(no one scored lower than 9)	
Glass jam jar	77%		
Paper napkin	73%		
Plastic bag	65%		
Gift-wrapping paper	63%		
Ceramic flowerpot	58%		
Wine bottle	45%		
Air freshener spray container	28%		

males tending to provide more correct answers, as well as income range (B = 0.145; p-value = 0.012), with higher-income households giving more correct answers. We also did a separate test for correlation between recycling proportion and revealed recycling knowledge, but the results were insignificant with a high p-value over 0.3, which suggests that recycling participation is unfortunately not associated with recycling knowledge. These results could inform targeted education interventions (i.e., discouraging wishful recycling, focusing on significant demographic groups), but the effect sizes for the significant independent variables were small. Hypothesis 1b was only partially supported. Overall, the analysis underscores the complexity of factors influencing recycling knowledge.

Qualitative comments on questions about recycling information reveal some additional context. Many respondents (38 comments) asked for information on which materials go into which bags, with quotes like this one: "A more complete list of recylables [*sic*]. Do glass wine bottles go into blue bags?" However, as noted above (and in two participant comments), such information is already publicly available, so there may be an issue of awareness and immediate accessibility (i.e., some residents will respond better to physical materials instead of online ones). There were also 21 comments asking for information improvements, such as detail

Table 10.3 Multiple linear regression results (correlation of eight independent variables with recycling knowledge as the dependent variable)

Coefficients	*Unstandardized coefficients*		*Standardized coefficients*	*t*	*Significance (p)*
Independent variables	*B*	*Standard error*	*Beta*		
(Constant)	17.437	1.182		14.750	<.001
Age	.065	.113	.040	.576	.565
Gender	–.371	.145	–.176	–2.555	.011*
Education	–.048	.174	–.020	–.278	.781
Income	.145	.057	.177	2.530	.012*
Availability of information	.115	.240	.052	.479	.633
Usefulness of information	.192	.242	.089	.795	.428
Clarity of labels	–.130	.177	–.059	–.736	.463
Wishful recycling	–.602	.178	–.234	–3.372	<.001*
Model summary	*R*	*R-squared*	*Adjusted R-squared*	*Standard error of estimate*	
	.331	.110	.073	2.310	

on what the different recycling numbers on plastic products mean or clearer visuals. The themes of limited accepted materials (18 comments) and distrust in the system (15 comments) appeared again, but others (two comments) believed some distrust was due to lack of awareness that there are two compartments in the garbage trucks (i.e., clear and blue bags are not being mixed together), a knowledge gap that could be clarified in educational materials. For instance, one participant said "explain to residents that there are 2 compartments in garbage truck [*sic*] (I hear lots of people say it goes in the same truck so why recycle)" (i.e., some people mistakenly think that the recycling gets mixed in with the garbage at the point of collection). Finally, there were six comments expressing a (perhaps mistaken) belief that the hauler will not pick up blue bags if they contain any incorrect items, corresponding to the wishful recycling motivation of some respondents.

Strategies to enhance recycling behavior (Research question 2)

Some areas for possible intervention were already identified in the above sections, but we also administered direct questions about management on the survey. One relevant question asked participants to indicate their level of satisfaction with various components of the recycling system in Western Newfoundland, on a Likert scale from "very dissatisfied" to "very satisfied". Table 10.4 is a descriptive summary of the ratio of satisfaction to dissatisfaction, with calculations and columns similar to Table 10.1. Respondents were most satisfied with convenience of the

Table 10.4 Respondent satisfaction with various components of recycling system

Component of recycling system	*Ratio of satisfaction to dissatisfaction across responses*	*% of participants indicating neither satisfaction nor dissatisfaction (or unsure/no answer)*
Convenience of blue-bag pick-up or drop-off	8.21	13%
Bag limits	2.78	25%
Convenience of drop-off at special collection sites	1.80	27%
Average quality of recycling services	1.65	32%
Accessibility and clarity of information on recycling	1.00	27%
Range of materials that can be recycled	0.83	22%
Garbage fees	0.75	35%
How proper sorting is enforced or explained	0.72	30%

blue-bag program (8.21 times as much satisfaction as dissatisfaction, or satisfaction ratio) and bag limits (2.78 satisfaction ratio). They were least satisfied with how proper sorting is enforced or explained (0.72 satisfaction ratio), garbage fees (0.75 satisfaction ratio), and range of materials accepted (0.83 satisfaction ratio); these components all triggered more dissatisfaction than satisfaction and are potential areas of intervention to increase resident support for the recycling system.

Another question with the potential to inform recycling improvements asked participants to predict how various recycling education or management strategies might affect them. We used a Likert scale from "would cause me to recycle much less/worse" to "would cause me to recycle much more/better". Using similar structure and assumptions to Tables 10.1 and 10.4, Table 10.5 provides a descriptive summary of the ratio of estimated positive effects (i.e., more/better recycling) to estimated negative effects (i.e., less/worse recycling). In general, the softer (typically educational) strategies were estimated to have a much stronger positive effect than the stricter (typically regulatory) strategies, but every strategy was estimated to have a positive effect, rather than a negative one, on average (i.e., every ratio was above 1). The strategies with the strongest estimated positive effect were information on what items go where (33.00 times as much positive effect estimation as negative effect estimation, or positive effect ratio) and information on how materials will be recycled (30.67 positive effect ratio), which aligns with the earlier reported confusion about where to put certain items (and errors on the quiz table) and distrust that the system is recycling materials effectively.

As for a relevant qualitative responses, we also asked respondents directly for their suggestions to improve the recycling system in Western Newfoundland. Emerging themes included: widen the range of materials (e.g., glass) accepted (25 comments); improve education, especially on what items can be taken where (14 comments); increase availability of drop-off sites and convenience of blue-bag

Table 10.5 Respondent predicted effects of education and management strategies in recycling system

Education or management strategy	*Ratio of estimated positive effects to estimated negative effects across responses*	*% of participants indicating neither positive nor negative effects (or unsure/no answer)*
Thorough educational materials on what items go where	33.00	24%
Thorough educational materials on how materials will be recycled	30.67	27%
Information on recycling/contamination rates in your community	23.33	43%
Educational notes provided for improperly sorted recycling/garbage	15.60	33%
Rewards/credits for properly sorted recycling	15.20	40%
High fines and enforcement for illegal dumping	5.33	49%
Clear (garbage) bags not picked up/accepted if they contain recyclables	1.46	37%
Blue (recycling) bags not picked up/accepted if they contain garbage	1.44	35%
Strict limits on number of clear (garbage) bags allowed	1.33	54%

program (11 comments); reduce fees at drop-off sites, especially to curb illegal dumping and burning (11 comments); and enforce stricter limits and fines on noncompliant residents (seven comments). More specifically, one participant said "We need to incorporate glass and styrofoam into our recycle [*sic*]. Also composting is crucial" (range of materials theme) and another stated "Make it more accessible, until they do ill [*sic*] continue to burn the excess rather than pay their outrageous tipping fees" (reduce fees theme). The focus on materials, education, and accessibility is not surprising given the responses already analyzed in earlier sections.

Discussion and conclusion

Comparing this case to other jurisdictions examined in the recycling literature, there are a few similarities across Miafodzyeva and Brandt's (2013) categorization of variables. Aligning with the inconsistent findings on demographic factors, we uncovered no evidence (yet) of demographic influences on recycling amount, and only slight effects of a few demographic variables (i.e., gender and income) on recycling knowledge. For sociopsychological variables, respondents self-reported the high positive influence of responsibility to future generations, conscientiousness, and environmental motivations, but we did not find statistical evidence of these factors, except for the minor effect of wishful recycling on recycling knowledge. The most salient category for Western Newfoundland, as suggested by our study, is technical-organization influences. Respondents routinely commented qualitatively on problems with the system and changes they would like to see (e.g., better sorting information, system transparency, broader education, range of materials, availability and convenience, lower fees, better enforcement), even to questions not directly asking about the system. While we did not acquire statistical evidence verifying the salience of these factors, our descriptive summaries and qualitative analysis suggest that they may be especially important for this region. In particular, the apparent distrust in the system is a factor not commonly reported for other jurisdictions. Thus, Western Newfoundland deserves unique recommendations for improving the blue-bag program (and other recycling programs).

Several points of intervention and possible strategies are identified by this study, but many would require extensive and perhaps controversial changes to the system (e.g., adding glass recycling, relaxed sorting and cleaning requirements). So, for practical reasons we identify two education-focused recommendations, which emerged across all our research questions:

a Improve available information on what items go where. Use clear visuals and various mediums (i.e., print and online) to increase accessibility. Focus on the most problematic items revealed by the recycling quiz. Clarify why certain items can or cannot be recycled. Address points of confusion like the recycling numbers on plastic items. Explain what happens when items are put in the wrong places. This recommendation is justified because survey respondents asked for it (implicitly in quantitative answers and explicitly in qualitative answers). Also, the negative influence of wishful recycling suggests that some residents have good intentions but need better guidance to ensure their actions are useful.

b Distribute transparent information on what happens to items after they are collected. Dispel common misconceptions (e.g., that clear and blue bags are mixed together in garbage trucks) with accurate information. Explain the process that materials go through to reach their ultimate destination. Quantify environmental and economic benefits and costs of the recycling system, demonstrating its value. This recommendation is justified because respondents asked for it (implicitly in quantitative answers and explicitly in qualitative answers). It would be a first step in improving public trust in the system, which the survey suggests is a root cause of negative perception and noncompliance.

Of course, this study was limited by its small sample size and imperfect representation of the population. Future studies on recycling in Western Newfoundland might improve the sample size (or at least sample a different segment of the population) with phone, mail, or door-to-door surveys. They might seek to test different hypotheses, collect different data, and ask questions in different ways (e.g., a more straightforward Likert scale from "strongly oppose" to "strongly support" for different management and education strategies). Still, this study provides insight into a region that has not been studied previously, discovering not only challenges but also a public with great potential to enthusiastically and effectively participate in recycling if the right factors can be addressed.

References

Anderson, D., & Krettenauer, T. (2021). Connectedness to nature and pro-environmental behaviour from early adolescence to adulthood: A comparison of urban and rural Canada. *Sustainability*, 13(1), 3655–3671.

Dahlén, L., & Lagerkvist, A. (2010). Pay as you throw: Strengths and weaknesses of weight-based billing in household waste collection systems in Sweden. *Waste Management*, 30(1), 23–31. https://doi.org/10.1016/j.wasman.2009.09.022.

DiGiacomo, A., Wu, D. W. L., Lenkic, P., Fraser, B., Zhao, J., & Kingstone, A. (2018). Convenience improves composting and recycling rates in high-density residential buildings. *Journal of Environmental Planning and Management*, 61(2), 309–331. https://doi.org/10.1080/09640568.2017.1315039.

Fan, B., Yang, W., & Shen, X. (2019). A comparison study of "motivation-intention-behavior" model on household solid waste sorting in China and Singapore. *Journal of Cleaner Energy Production*, 20(1), 442–454.

Ferrara, I., & Missios, P. (2005). Recycling and waste diversion effectiveness: Evidence from Canada. *Environmental and Resource Economics*, 30(2), 221–238. https://doi.org/10.1007/s10640-004-1518-z.

Fracassi, J. (2022, March). *Assessing the waste management system in the Gros Morne region*. AHOI. www.ahoi.ca/_files/ugd/4108a4_628dce24df994a849a0c92a15ac9b111.pdf (accessed June 18, 2024).

Gifford, R., & Nilsson, A. (2014). Personal and social factors that influence pro-environmental concern and behaviour: A review. *International Journal of Psychology*, 49(3), 141–157. https://doi.org/10.1002/ijop.12034.

Guiot, D., Malas, Z., & Urien, B. (2019). Re-examining the effects of aging on household recycling behavior: A study in France. *Psychology & Marketing*, 36(6), 618–629. https://doi.org/10.1002/mar.21200.

Hoffmann, R., Muttarak, R., Peisker, J., & Stanig, P. (2022). Climate change experiences raise environmental concerns and promote green voting. *Nature Climate Change*, 12(2), 148–155. https://doi.org/10.1038/s41558-021-01263-8.

Kennedy, E. H., & Kmec, J. (2018). Reinterpreting the gender gap in household pro-environmental behaviour. *Environmental Sociology*, 4(3) 299–310. https://doi.org/10.1080/23251042.2018.1436891.

Krettenauer, T., & Lefebvre, J. P. (2021). Beyond subjective and personal: Endorsing pro-environmental norms as moral norms. *Journal of Environmental Psychology*, 76, 101644. https://doi.org/10.1016/j.jenvp.2021.101644.

Lakhan, C. (2015). Evaluating the effects of unit-based waste disposal schemes on the collection of household recyclables in Ontario, Canada. *Resources, Conservation and Recycling*, 95, 38–45. https://doi.org/10.1016/j.resconrec.2014.12.002.

Lee, E. Y., & Khan, A. (2020). Prevalence and clustering patterns of pro-environmental behaviors among Canadian households in the era of climate change. *Sustainability*, 12(19), 8218. https://doi.org/10.3390/su12198218.

Miafodzyeva, S. & Brandt, N. (2013). Recycling behaviour among householders: Synthesizing determinants via a meta-analysis. *Waste Biomass Valor*, 4(1), 221–235. https://doi.org/10.1007/s12649-012-9144-4.

Multi-Material Stewardship Board. (2019). *Waste audit report: Western NL curbside waste*. ww.wrwm.ca/wp-content/uploads/2015/10/WRWM-Waste-Audit-Report-FINAL.pdf (accessed June 12, 2024).

Munro, A. (2010). *Recycling by Canadian households, 2007. Environment accounts and statistics analytical and technical paper series*. www150.statcan.gc.ca/n1/pub/16-001-m/16-001-m2023001-eng.pdf (accessed June 9, 2024)

Qualtrics. (2024). *Sample size calculator*. www.qualtrics.com/blog/calculating-sample-size/ (accessed August 27, 2024).

Rosenthal, S., & Linder, N. (2021). Effects of bin proximity and informational prompts on recycling and contamination. *Resources, Conservation and Recycling*, 168, 105430. https://doi.org/10.1016/j.resconrec.2021.105430.

Smith, C. J., Dupré, K. E., McEvoy, A., & Kenny, S. (2021). Community perceptions and pro-environmental behavior: The mediating roles of social norms and climate change risk. *Canadian Journal of Behavioural Science/Revue canadienne des sciences du comportement*, 53(2), 200. https://doi.org/10.1037/cbs0000239.

Statistics Canada. (2023). *Profile table, census profile, 2021 census of population—West Coast—Northern Peninsula—Labrador (Economic region), Newfoundland and Labrador*. www12.statcan.gc.ca/census-recensement/2021/dppd/prof/details/page.cfm?Lang=E&SearchText=West%20Coast%2D%2DNorthern%20Peninsula%2D%2DLabrador&DGUIDlist=2021S05001030&GENDERlist=1,2,3&STATISTIClist=1,4&HEADERlist=0 (accessed June 7, 2024).

Stojanovski, O., Leslie, G. W., Wolak, F. A., Huerta Wong, J. E., & Thurber, M. C. (2020). Increasing the energy cognizance of electricity consumers in Mexico: Results from a field experiment. *Journal of Environmental Economics and Management*, 102, 102323. https://doi.org/10.1016/j.jeem.2020.102323.

Viscusi, K., Huber, J., & Bell, J. (2014). Private recycling values, social norms, and legal rules. *Revue d'économie politique*, 124(2), 159–178.

Western Regional Waste Management. (2019). *The industrial, commercial, and institutional (ICI) policy*. www.wrwm.ca/wp-content/uploads/2015/10/ICI-Waste-Policy-Final-Oct-28-19.pdf (accessed June 18, 2024).

Western Regional Waste Management. (n.d.). *Waste disposal sites*. www.wrwm.ca/services/waste-disposal-sites/ (accessed June 14, 2024).

Yang, C., & Arhonditsis, G. B. (2022). What are the primary covariates of environmental attitudes and behaviours in Canada? A national-scale analysis of socioeconomic, political, and demographic factors. *Ecological Informatics*, 69, 101661. https://doi.org/10.1016/j.ecoinf.2022.101661.

Part 4

Landfills and innovative waste management

11 Field investigation and numerical modeling of MSW settlement in cold regions

Wameed Alghazali, Paul J. Van Geel and Shawn Kenny

Introduction

In cold regions such as Canada, municipal solid waste (MSW) is frequently exposed to freezing conditions prior to landfill disposal, which can impact settlement and biodegradation. MSW can remain partially frozen for months or even years due to the effects of latent heat and insulation from overlying waste layers (Alghazali et al., 2024a; Berquist & Van Geel, 2020; Hanson et al., 2010; Van Geel & Murray, 2015). This extended freezing period delays waste biodegradation-induced settlement, which is one of the three main components contributing to landfill settlement. The two other components of MSW settlement are immediate or instantaneous settlement, resulting from the overburden pressure caused by upper waste layers, and mechanical creep, occurring over time under constant overburden pressure (Alghazali et al., 2024a).

MSW settlement is critical to optimizing the use of landfill airspace and overall site performance (Alghazali et al., 2024b; Ren et al., 2022). Research on MSW settlement has been conducted at various scales to develop a better understanding of its behavior. Valuable insights have been gained from small-scale (Bareither et al., 2012a, 2013; Basha et al., 2016; Hossain et al., 2003; Mousavi, 2021), large-scale (Bareither et al., 2012b; Mousavi, 2021), and landfill instrumentation (Jo & Jang, 2021; Mehta et al., 2002; Simões & Catapreta, 2013) studies. A wide range of compressibility values have been observed, depending on factors such as sample composition, size, and type (e.g., field versus synthetic samples, age, degree of shredding, etc.). It is important to note that, with the exception of a study conducted at the Sainte Sophie Landfill in Québec, Canada (Alghazali et al., 2024a), laboratory and field research has not specifically examined the effects of freezing temperatures on MSW stabilization. This chapter, therefore, will focus on the Sainte Sophie Landfill study, which provided long-term settlement and temperature data from lifts placed under both freezing and warm ambient temperatures. This unique data set will help us understand the impact of the low temperatures experienced in cold regions on MSW biodegradation-induced settlement.

Landfill settlement is typically estimated using numerical models, and several types have been proposed in the literature to simulate MSW settlement.

DOI: 10.4324/9781003534129-16

Review papers (El-Fadel & Khoury, 2000; Lu & Feng, 2020; Ren et al., 2022) have categorized numerical settlement models into soil mechanics based models (Hossain & Gabr, 2005; Reddy et al., 2018; Shi et al., 2016; Sowers, 1973), rheological models (Alghazali et al., 2024a; Gibson & Lo, 1961), empirical models (Edil et al., 1990; Ling et al., 1998; Yen & Scanlon, 1975), biodegradation-induced settlement models (Berquist & Van Geel, 2020; Marques et al., 2003; Park & Lee, 1997), and constitutive models (Feng et al., 2016; Gao & Kavazanjian, 2022; Kumar et al., 2021; Machado et al., 2009; McDougall, 2007; Mousavi & Eun, 2023; Sivakumar Babu et al., 2010). A recent comprehensive assessment of the effectiveness of the most common settlement models, particularly at the field scale, is presented by Alghazali et al. (2024c).

In addition to the settlement and temperature data collected from the Sainte Sophie Landfill, the authors have developed a thermal-mechanical-biological (TMB) model designed to simulate both MSW temperature and settlement (Alghazali et al., 2024a). The model relies on simple input parameters, which can be obtained either from field measurements or laboratory experiments. It incorporates temperature-driven biodegradation processes, making it well suited for simulations in cold climate regions. The model's formulation and its application will be examined later in this chapter, following the discussion of the field data.

Landfill instrumentation and field data

Site instrumentation

Sainte Sophie Landfill, located about 50 km northwest of Montreal, Québec, Canada, is equipped with a leachate collection system and horizontal gas collection wells. To study waste stabilization processes in cold climates, 12 monitoring bundles were installed in two vertical columns at different depths during the landfill's filling operations as illustrated in Figure 11.1. The instrumentation bundles were equipped with temperature sensors, a liquid settlement system, a total earth pressure cell (TEPC), an oxygen sensor, a moisture and electrical conductivity sensor, and a vibrating wire piezometer. The liquid settlement system features a vibrating wire pressure transducer linked hydraulically to a reference transducer, which is located in an instrument shed outside the landfill footprint. To prevent freezing, the hydraulic connection uses tubing filled with a solution of 50% water and 50% glycol. Upon installation, an initial reference differential pressure was recorded. As settlement occurs, the transducers measure corresponding changes in differential pressure. The pressure changes are then converted into settlement data, adjusted for liquid density and temperature variations. For more detailed information on the instrumentation, refer to the studies by Megalla et al. (2016) and Vingerhoeds (2011).

The monitored area of the landfill includes six waste lifts and a final cover. The initial three waste lifts were placed during the winter season, when freezing conditions prevailed, whereas the remaining lifts were deposited during the fall and summer under significantly warmer temperatures. The waste materials were

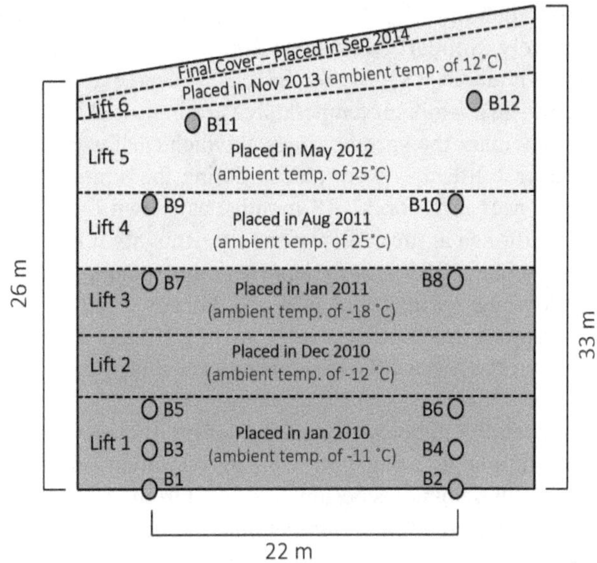

Figure 11.1 Cross-section of the instrumented area of the Sainte Sophie Landfill.

Notes: "B" denotes Bundle (e.g., B1 refers to Instrument Bundle 1).

arranged in layers, each approximately 0.3–0.4 m thick, and were compacted using four to five passes of a sheepsfoot roller to ensure density and stability. Multiple compacted layers combine to form a lift, which was then covered with an interim cover until the next lift was added. At the Sainte Sophie site, the disposed MSW was composed of 53% residential waste, 8% commercial waste, 10% sludge along with construction and demolition debris, and 29% soil, which was used as daily cover (Berquist & Van Geel, 2020). The compacted waste at the site achieved an average density of roughly 930 kg/m³ (Alghazali et al., 2024a). Notably, the initial waste layer had a thickness of 1.5 m and was placed as a single uncompacted layer above the leachate collection system. Further details regarding the placement times, thickness of the initial waste lifts, and the specific locations of the monitoring bundles are provided by Alghazali et al. (2024a).

Settlement and temperature data

Data from the 12 monitoring bundles, including both settlement and temperature readings, were continuously recorded every 30 minutes over a period of more than 12 years. This chapter presents field data related to settlement and temperature from the lower monitoring bundles installed in the lifts placed during the winter months (lifts 1–3), as illustrated in Figure 11.2. In this chapter, settlement data corresponding to each bundle is labeled with the letter "S" followed by the specific bundle number. Temperature data, on the other hand, is denoted by the letter "T" alongside the relevant bundle number. Figure 11.2(a, b) displays the temperature data from the odd-numbered bundles (T1, T3, T5, T7) and even-numbered bundles (T2, T4, T6, T8), respectively. Similarly, Figure 11.2(c, d) shows the settlement data from the odd-numbered bundles (S3–S1, S5–S3, S5–S1) and even-numbered bundles (S4–S2, S6–S4, S6–S2). Ambient temperatures are indicated by the light gray line, while vertical bars mark the specific times at which each waste lift was deposited in the field. Because lifts 1–3 were placed during the winter, the temperature in lift 1 remained near zero for 12–18 months, as shown by data from T3 and T4, which were positioned at the center of the lift. Roughly 4 m of waste, corresponding to the second half of lift 1, was deposited above the bundles, providing insulation. This, combined with latent heat effects, prevented significant temperature changes in response to summer conditions in 2010. As a result, the onset of biological activity in the lower lifts was delayed, causing a slower rise in temperatures over time.

Bundles B5 and B6, placed 0.5 m below the surface of lift 1, closely followed ambient temperatures with a slight phase lag. Following the placement of the second waste lift in December 2010, these bundles became insulated from the surrounding atmosphere and no longer reflected ambient temperature variations. Bundles B7 and B8, located 0.5 m below the surface of lift 3, recorded increasing temperatures between January 2011 and August 2011, following ambient temperatures with a slight phase lag. These bundles were then insulated from the atmosphere when lift 4 was placed in August 2011. The initial temperature of the waste in lift 4 was 25.2°C, and the temperatures recorded by T7 and T8 gradually increased in response to the heat from the overlying warmer lift and the heat generated by waste biodegradation within lift 4.

In January 2013, the temperature in lift 1 began to gradually rise (T3 and T4), reaching 10–12°C by October 2013. At this point, biodegradation-induced settlement commenced in lift 1, as indicated by a new slope in the settlement data over time (S3–S1 and S4–S2). The base of the landfill is underlain by a clay layer, which consolidated over time due to the increased overburden pressure from waste placement. To isolate the settlement occurring in the waste lifts only, the settlement data recorded at bundles B1 and B2, located at the base of the landfill, were subtracted from the data of the odd and even bundles, respectively. As shown in Figure 11.2(c, d), each odd and even pair of instrument bundles (S3–S1 and S4–S2, S5–S3 and S6–S4) followed similar settlement paths over time, which reinforces the reliability of the collected settlement data.

The settlement data displayed a clear instantaneous response due to the placement of the lifts above the bundles, which was relatively small compared to the total settlement. The magnitude of this instantaneous response varied and decreased as the overburden pressure above lift 1 increased. For instance, the instantaneous settlement response to the placement of lift 4 was significantly less than that observed due to the placement of the second half of lift 1. Figure 11.2 (c, d) also suggests that the mechanical creep experienced by MSW depends on the applied stress. For example, S3–S1 and S4–S2 experienced less mechanical creep when lifts 2 and 3 were placed compared to the mechanical creep observed

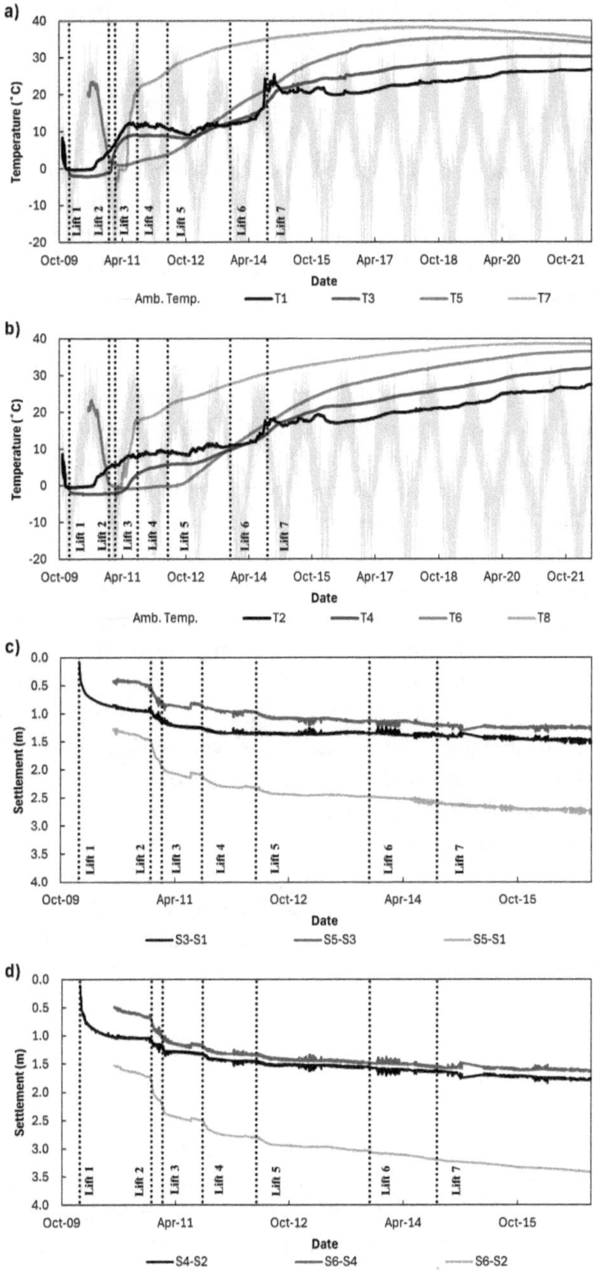

Figure 11.2 Field data collected from the Sainte Sophie Landfill. (a) Temperatures at the odd bundles; (b) Temperatures at the even bundles; (c) Settlements at the odd bundles; (d) Settlements at the even bundles.

when the second half of lift 1 was placed. Similar findings regarding the increase in MSW stiffness as a result of increased applied stress have been reported in other studies (Beaven & Powrie, 1995; Kumar et al., 2021; Landva et al., 2000).

Landfill modeling

In the literature, MSW settlement is categorized into three components: instantaneous settlement, mechanical creep, and biodegradation-induced settlement. Each component has been estimated using various models. Instantaneous settlement prediction methods include the compression ratio (Li et al., 2021; Marques et al., 2003; Sowers, 1973), elastic models with yield criteria (Reddy et al., 2018), and the Modified Cam-Clay model (Machado et al., 2009; McDougall, 2007; Mousavi & Eun, 2023). Time-dependent mechanical creep is analyzed using compression ratios (Sowers, 1973), rheological models (Gibson & Lo, 1961; McDougall, 2007; Mousavi & Eun, 2023), the equivalent time model (McDougall, 2007), the Soft-Soil-Creep model (Kumar & Reddy, 2021), and first-order kinetic formulas (Chen et al., 2010). Models for biodegradation-induced settlement include the compression ratio (Hossain & Gabr, 2005), first-order kinetic formulas (Gao & Kavazanjian, 2022; Mousavi & Eun, 2023), degradation-induced void change parameter (Kumar & Reddy, 2021), and empirical functions (Berquist & Van Geel, 2020).

The authors proposed a framework for a TMB model that requires a limited number of input parameters, which can be evaluated from either simple laboratory experiments or field data from instrumented landfills (Alghazali et al., 2024a). This model was designed to simulate the settlement and temperature of MSW in landfills, including landfills in cold climates. The model was developed using COMSOL Multiphysics and validated and calibrated using the 12 years of field data collected from the Sainte Sophie Landfill. To provide readers with an overview of the TMB model, its formulation and application will be briefly discussed in hereafter.

The TMB model represents heat transfer through conduction, as described in Equation 11.1. The model adjusts the thermal conductivity and volumetric heat capacity, the latter being the product of specific heat capacity and density, based on continuously updated density values due to waste settlement.

$$\nabla(k\nabla T)+Q=\rho_{updated}\,C_p\,\frac{\partial T}{\partial t} \tag{11.1}$$

In this equation, k denotes thermal conductivity, T stands for temperature, C_p represents specific heat capacity, $\rho_{updated}$ is the updated density calculated through the model's mechanical component, and Q is a source/sink term accounting for heat loss and gain at the lift surfaces due to radiation and convection when exposed to the atmosphere. Q also incorporates heat generated within the lifts due to biological processes, based on a temperature-dependent heat generation model proposed by Hanson et al. (2013).

Latent heat effects are also integrated into the TMB model through adjustments to specific heat capacity, as shown in Equation 11.2. When MSW temperature ranges from −1.6 to −0.6°C, specific heat capacity (C_p) is modified by a latent heat term (L), accounting for the energy required to thaw the liquid fraction of frozen MSW, assumed to occur over a temperature change of 1°C (Bonany et al., 2013). Detailed description of the thermal model is available in Alghazali et al. (2024b).

$$C_p = \frac{L}{T} \text{ when } -1.6°\text{C} < T < -0.6°\text{C} \tag{11.2}$$

The TMB model incorporates a Generalized Kelvin-Voigt (GKV) model, which features a spring connected in series with a Kelvin-Voigt (KV) body (Figure 11.3), to define the stress–strain relationship. The stress in the elastic branch matches that in the KV body, with the elastic strain being the difference between total and viscoelastic strains. The model expresses the modulus of elasticity as a function of overburden stress using a power function (Janbu, 1963). Key input parameters for the GKV model include the modulus of elasticity for the springs, Poisson's ratio for the springs, and a relaxation time for the dashpot.

Figure 11.4 depicts a typical gas and corresponding heat generation rate over time at two different temperatures. Integrating temperature into the model complicates tracking heat generation rates, as transitioning between curves requires temporal adjustments. Hanson et al. (2013) suggested expressing heat generation

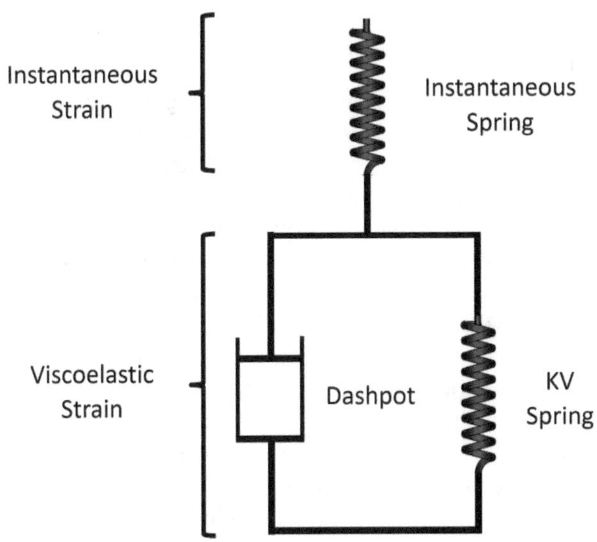

Figure 11.3 Illustration of the GKV model.

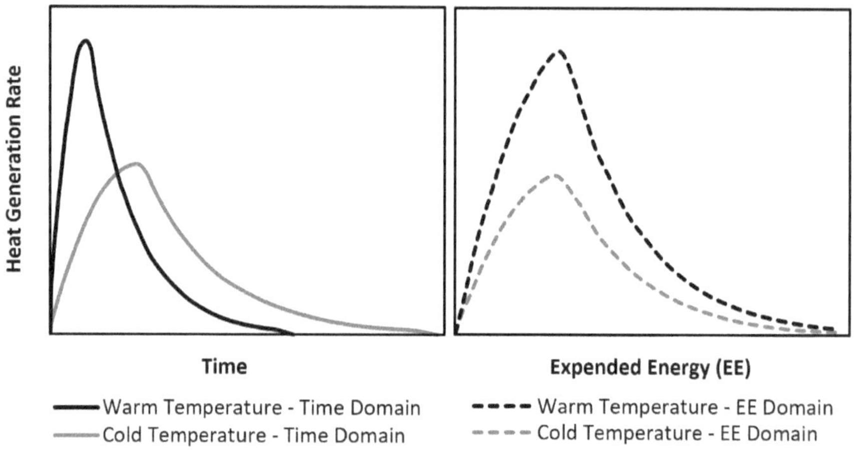

Figure 11.4 Heat generation curves for two different temperatures.

rate as a function of expended energy, enabling the model to use expended energy and temperature to predict heat generation rates (Figure 11.4). This approach was adopted in the TMB model to simulate heat generation from MSW and estimates biodegradation-induced settlement based on waste expended energy. Further detailed information on the TMB formulation, along with a list of calibrated input parameters, can be found in Alghazali et al. (2024a) and Alghazali et al. (2024b).

TMB modeling results

The TMB model's simulated temperatures are illustrated in Figure 11.5. The model captured the delayed temperature rise observed in the lower bundles (T3, T4, T5, T6) due to the placement of MSW under freezing conditions. Additionally, the TMB model showed that bundles positioned near the surface of the lift, such as T5 and T6 placed at the top of lift 1 or T7 and T8 placed at the top of lift 3, followed ambient temperatures closely with a slight phase lag until the next lift was added. Following the placement of lift 4, the simulated temperatures at T7 and T8 in lift 3 initially showed a decline, whereas the field data indicated a slight increase in temperatures. Reviewing the simulated vertical temperature profiles during this period revealed a sharp initial temperature contrast between lifts 3 and 4 at the time of placement of lift 4. The model tended to smooth this sharp contrast to generate a more uniform downward thermal gradient, leading to the initial slight temperature decline at T7 and T8. Lift 3 was placed during winter, and when lift 4 was added, the center of lift 3 remained frozen based on field temperatures recorded by bundles B5 and B6, while the initial temperature of lift 4 was 25°C (Alghazali et al., 2024b).

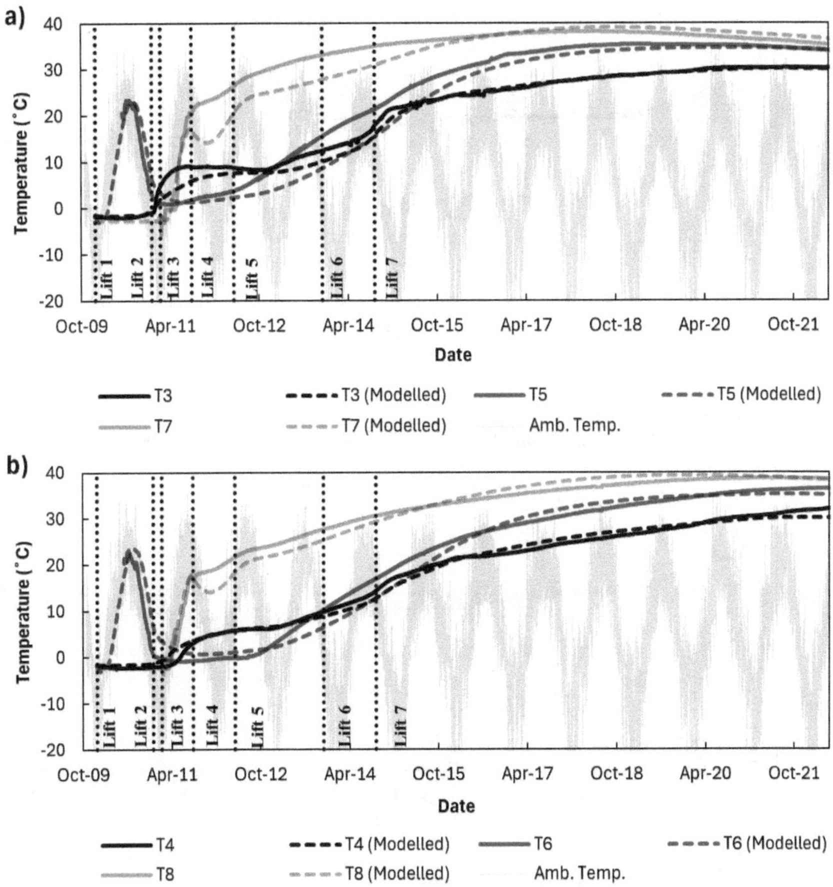

Figure 11.5 Field and modeled temperatures. (a) Odd bundles; (b) Even bundles.

Figure 11.6 compares the field and modeled settlements at the odd and even bundles. The model captured the resistance to compressibility in MSW, reflected in instantaneous and mechanical creep settlements upon the addition of new lifts. To highlight the model's capability to simulate the delay in biodegradation-induced settlement observed in the field due to the placement of MSW under freezing conditions, the simulated settlements with and without biodegradation-induced settlement for S3–S1, S4–S2, S5–S1, and S6–S2 are plotted in Figure 11.6. These results demonstrate the model's capability in simulating the biological and thermal processes of MSW by predicting temperatures and biodegradation-induced settlements over time. For instance, during the early period from January 2010 to October 2013, when the temperature of lift 1 was

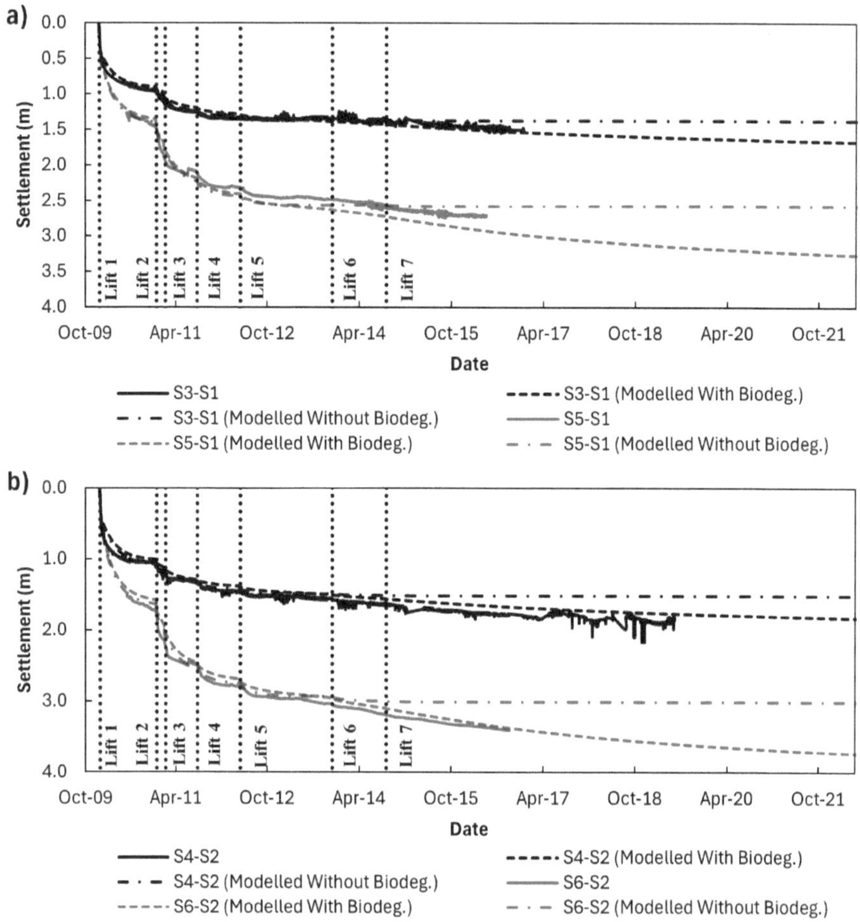

Figure 11.6 Field and modeled settlements. (a) Odd bundles; (b) Even bundles.

not conducive to biodegradation, the model predicted negligible biodegradation-induced settlement. Consequently, the simulated settlement was primarily instantaneous settlement and mechanical creep. However, deviations between simulated settlements with and without biodegradation-induced settlement became evident around October 2013, marking the onset of temperatures favorable for MSW biodegradation. The model simulated the increased rate of settlement observed in the field around October 2013 in response to rising waste temperatures, underscoring the importance of incorporating a thermal model and temperature-dependent biodegradation in simulations of landfill settlement in cold regions (Alghazali et al., 2024a).

Study implications

A critical requirement for the design of any landfill is the accurate estimation of its airspace capacity, which depends on the overall landfill settlement. This settlement includes instantaneous settlement, mechanical creep, and biodegradation-induced settlements. One significant implication of the present study is the application of the TMB model to estimate MSW settlement under various environmental and operational conditions. The field data collected in this study highlights that freezing temperatures in cold regions hinder waste biodegradation, consequently delaying the biodegradation-induced settlement component. This finding is crucial for landfill operations in cold regions, such as those in Canada. The TMB model and field data presented in this study can assist practitioners in these regions to optimize landfill design and operational strategies to mitigate the adverse effects of cold temperatures on waste stabilization. For instance, the TMB model can be used to simulate different waste placement scenarios under varying ambient temperatures. Practitioners can analyze the impact of these scenarios on settlement, particularly when subsequent waste lifts are placed during winter months. By doing so, they can plan the timing and sequence of waste placement to enhance biodegradation-induced settlement, thereby improving the overall efficiency and capacity of the landfill. Moreover, the insights gained from this study can inform the development of guidelines for landfill operation in cold climates. By integrating these findings into landfill design and management practices, practitioners can achieve more sustainable and efficient landfill operations, ensuring better long-term performance.

Conclusion

Understanding the thermal dynamics and settlement behavior of MSW in cold regions is crucial for optimizing landfill design and operation, particularly in terms of airspace management. This chapter presents a comprehensive field investigation and numerical modeling study of MSW stabilization in cold regions, focusing on the Sainte Sophie Landfill in Québec, Canada. It also discusses a previously developed TMB model capable of simulating MSW settlements and temperatures in landfills.

The collected field data spanned over 12 years and provided valuable insights into the behavior of MSW under varying temperature conditions. They revealed significant delays in temperature rise within the lower waste lifts, which remained partially frozen for extended periods due to initial placement under freezing conditions. This prolonged freezing period delayed the onset of waste biodegradation and affected waste settlement rates. Additionally, the settlement data indicated a clear resistance to compressibility in the MSW under increasing overburden pressure, which is important for estimating a landfill's airspace capacity.

The proposed TMB model was employed to simulate the settlement and temperatures at the Sainte Sophie Landfill. The model successfully captured the observed temperature and settlement behaviors in the field, including the delayed temperature rise and subsequent biodegradation-induced settlement. Its ability to account for temperature-dependent biodegradation processes makes it particularly suitable for landfills in cold regions. The findings emphasize the need to account for temperature variations in MSW landfill models, particularly in cold regions where freezing can delay settlement.

References

Alghazali, W., Kenny, S., & Van Geel, P. J. (2024a). A comparative study of models simulating the settlement of municipal solid waste (MSW) in landfills. *Waste Management*. Waste Management. 190, 236–250. https://doi.org/10.1016/j.wasman.2024.09.024

Alghazali, W., Van Geel, P. J., & Kenny, S. (2024b). A comprehensive study of settlement during the filling and post- closure phases at a landfill in Québec, Canada: Field data and TMB modelling. *Waste Management*, 178, 85–96. https://doi.org/10.1016/j.wasman.2024.02.019

Alghazali, W., Kenny, S., & Van Geel, P. J. (2024c). A comparative study of models simulating the settlement of municipal solid waste (MSW) in landfills. *Waste Management*, 190, 236–250. https://doi.org/10.1016/j.wasman.2024.09.024

Bareither, C. A., Benson, C. H., & Edil, T. B. (2012a). Compression behavior of municipal solid waste: Immediate compression. *Journal of Geotechnical and Geoenvironmental Engineering*, 138(9), 1047–1062.

Bareither, C. A., Benson, C. H., & Edil, T. B. (2013). Compression of municipal solid waste in bioreactor landfills: Mechanical creep and biocompression. *Journal of Geotechnical and Geoenvironmental Engineering*, 139(7), 1007–1021.

Bareither, C. A., Breitmeyer, R. J., Benson, C. H., Barlaz, M. A., & Edil, T. B. (2012b). Deer track bioreactor experiment: field-scale evaluation of municipal solid waste bioreactor performance. *Journal of Geotechnical and Geoenvironmental Engineering*, 138(6), 658–670.

Basha, B. M., Parakalla, N., & Reddy, K. R. (2016). Experimental and statistical evaluation of compressibility of fresh and landfilled municipal solid waste under elevated moisture contents. *International Journal of Geotechnical Engineering*, 10(1), 86–98.

Beaven, R. P., & Powrie, W. (1995). Hydrogeological and geotechnical properties of refuse using a large scale compression cell. In *5th international waste management and landfill symposium* (pp. 745–760). Sardinia.

Berquist, C., & Van Geel, P. J. (2020). Simulating temperature-dependent biodegradation-induced settlement at a landfill with waste lifts placed under frozen conditions. *Waste Management*, 104, 74–81.

Bonany, J. E., Van Geel, P. J., Gunay, H. B., & Isgor, O. B. (2013). Heat budget for a waste lift placed under freezing conditions at a landfill operated in a northern climate. *Waste Management*, 33(5), 1215–1228.

Chen, Y., Ke, H., Fredlund, D. G., Zhan, L., & Xie, Y. (2010). Secondary compression of municipal solid wastes and a compression model for predicting settlement of municipal solid waste landfills. *Journal of Geotechnical and Geoenvironmental Engineering*, 136(5), 706–717.

Edil, T. B., Ranguette, V. J., & Wuellner, W. W. (1990). Settlement of municipal refuse. *Geotechnics of Waste Fills: Theory and Practice*, 225–239.

El-Fadel, M., & Khoury, R. (2000). Modeling settlement in MSW landfills: A critical review. *Critical Reviews in Environmental Science and Technology*, 30(3), 327–361.

Feng, S.J., Cao, B.Y., Bai, Z.B., & Yin, Z.Y. (2016). Constitutive model for municipal solid waste considering the effect of biodegradation. *Géotechnique Letters*, 6, 244–249. https://doi.org/ 10.1680/jgele.15.00184.

Gao, W., & Kavazanjian, E. (2022). A constitutive model for municipal solid waste considering mechanical creep and biodegradation-induced compression. *Acta Geotechnica*, 17(1), 37–63.

Gibson, R. E., & Lo, K. Y. (1961). *A theory of consolidation for soils exhibiting secondary compression*. Norwegian Geotechnical Institute.

Hanson, J. L., Yeşiller, N., & Oettle, N. K. (2010). Spatial and temporal temperature distributions in municipal solid waste landfills. *Journal of Environmental Engineering*, 136(8), 804–814.

Hanson, J. L., Yeşiller, N., Onnen, M. T., Liu, W. L., Oettle, N. K., & Marinos, J. A. (2013). Development of numerical model for predicting heat generation and temperatures in MSW landfills. *Waste Management*, 33(10), 1993–2000.

Hossain, M. S., & Gabr, M. A. (2005). Prediction of municipal solid waste landfill settlement with leachate recirculation. *Waste Contaminant and Remediation*, 2, 1–5.

Hossain, M. S., Gabr, M. A., & Barlaz, M. A. (2003). Relationship of compressibility parameters to municipal solid waste decomposition. *Journal of Geotechnical and Geoenvironmental Engineering*, 129(12), 1151–1158.

Janbu, N. (1963). Soil compressibility as determined by oedometer and triaxial tests. In proceedings of the *European conference on soil mechanics and foundation engineering (ECSMFE)* (Vol. 1; pp. 19–25). Wiesbaden, Germany.

Jo, Y.-S., & Jang, Y.-S. (2021). Comparison of waste settlement characteristics for two landfills disposed in long sequential periods. *Waste Management*, 131, 433–442.

Kumar, G., & Reddy, K. R. (2021). Comprehensive coupled thermo-hydro-bio-mechanical model for holistic performance assessment of municipal solid waste landfills. *Computers and Geotechnics*, 132(June 2020), 103920. https://doi.org/10.1016/j.compgeo.2020.103920

Kumar, G., Reddy, K. R., & Foster, C. (2021). Modeling elasto-visco-bio-plastic mechanical behavior of municipal solid waste in landfills. *Acta Geotechnica*, 16(4), 1061–1081.

Landva, A. O., Valsangkar, A. J., & Pelkey, S. G. (2000). Lateral earth pressure at rest and compressibility of municipal solid waste. *Canadian Geotechnical Journal*, 37(6), 1157–1165.

Li, K., Chen, Y. M., Xu, W. J., Zhan, L. T., Ling, D. S., Ke, H., Hu, J., et al. (2021). A thermo-hydro-mechanical-biochemical coupled model for landfilled municipal solid waste. *Computers and Geotechnics*, 134, 104090.

Ling, H. I., Leshchinsky, D., Mohri, Y., & Kawabata, T. (1998). Estimation of municipal solid waste landfill settlement. *Journal of Geotechnical and Geoenvironmental Engineering*, 124(1), 21–28.

Lu, S.-F., & Feng, S.-J. (2020). Comprehensive overview of numerical modeling of coupled landfill processes. *Waste Management*, 118, 161–179.

Machado, S. L., Carvalho, M. D. F., & Vilar, O. M. (2009). Modeling the influence of biodegradation on sanitary landfill settlements. *Soils and Rocks*, 32(3), 123–134.

Marques, A. C. M., Filz, G. M., & Vilar, O. M. (2003). Composite compressibility model for municipal solid waste. *Journal of Geotechnical Geoenvironmental Engineering*, 129(April), 372–378.

McDougall, J. (2007). A hydro-bio-mechanical model for settlement and other behaviour in landfilled waste. *Computers and Geotechnics*, 34(4), 229–246.

Megalla, D., Van Geel, P. J., & Doyle, J. T. (2016). Simulating the heat budget for waste as it is placed within a landfill operating in a northern climate. *Waste Management*, 55, 108–117.

Mehta, R., Barlaz, M. A., Yazdani, R., Augenstein, D., Bryars, M., & Sinderson, L. (2002). Refuse decomposition in the presence and absence of leachate recirculation. *Journal of Environmental Engineering*, 128(3), 228–236.

Mousavi, M. S. (2021). Multi-physical analysis of thermal, hydraulic, mechanical, and bio-chemical (THMB/C) processes in a municipal solid waste landfill. University of Nebraska-Lincoln.

Mousavi, M. S., & Eun, J. (2023). A predictive settlement modeling framework employing thermal-hydraulic-mechanical-biochemical processes in municipal solid waste landfills. *International Journal of Geomechanics*, *23*(6), 1–14.

Park, H. I., & Lee, S. R. (1997). Long-term settlement behavior of landfills with refuse decomposition. *Solid Waste Technology and Management*, 24(4), 159–165.

Reddy, K. R., Kumar, G., & Giri, R. K. (2018). Modeling coupled hydro-bio-mechanical processes in bioreactor landfills: Framework and validation. *International Journal of Geomechanics*, 18(9), 1–20.

Ren, Y., Zhang, Z., & Huang, M. (2022). A review on settlement models of municipal solid waste landfills. *Waste Management*, 149(May), 79–95.

Shi, J., Qian, X., Liu, X., Sun, L., & Liao, Z. (2016). The behavior of compression and degradation for municipal solid waste and combined settlement calculation method. *Waste Management*, 55, 154–164.

Simões, G. F., & Catapreta, C. A. A. (2013). Monitoring and modeling of long-term settlements of an experimental landfill in Brazil. *Waste Management*, 33(2), 420–430.

Sivakumar Babu, G. L., Reddy, K. R., & Chouksey, S.K. (2010). Constitutive model for municipal solid waste incorporating mechanical creep and biodegradation-induced compression. *Waste Management*, 30, 11–22. https://doi.org/10.1016/j.wasman.2009.09.005

Sowers, G. F. (1973). Settlement of waste disposal fills. In *8th international conference on soil mechanics and foundation engineering* (pp. 207–210). Balkema.

Van Geel, P. J., & Murray, K. E. (2015). Simulating settlement during waste placement at a landfill with waste lifts placed under frozen conditions. *Waste Management*, 46, 352–361. https://doi.org/10.1016/j.wasman.2015.08.016

Vingerhoeds, E. (2011). Instrument installation and preliminary data in a full-scale anaerobic bioreactor landfill, Sainte-Sophie, Quebec, Canada. Carleton University.

Yen, B. C., & Scanlon, B. S. (1975). Sanitary landfill settlement rates. *Journal of Geotechnical Engineering*, 101(5), 475–487.

12 Landfill biocell technology for northern climates

The Calgary Biocell

Poornima Jayasinghe, J. Patrick Hettiaratchi, Saranga Munasinghage, Seniru Ruwanpura, Hiroshan Hettiarachchi, Dinesh Pokhrel, Gopal Achari and Olufemi Akintunde

Introduction

The current global production rate of municipal solid waste (MSW) is about 2 billion tons annually, out of which more than 90% ends up in landfills (Gao et al., 2015). Until a few decades ago in Canada, and currently in most developing countries, open dumping on land was the most common approach to MSW management (Zohoori & Ghani, 2017). Research has shown that open dumps generate negative human health and environmental impacts (Siddiqua et al., 2022). Therefore, in Canada, open dumps have been primarily replaced with "dry-tomb type" sanitary landfills which are designed and constructed to eliminate human health and environmental problems associated with open dumps, such as contamination of groundwater by landfill leachate (Hettiaratchi, 2007).

However, conventional landfilling largely ignores sustainable development principles and faces several problems, including issues with landfill gas, long-term liability, and valuable space loss. In major cities, the rapid depletion of space for new landfills forces municipalities to seek new locations frequently, which is increasingly difficult due to public disinterest and limited real estate (Meegoda et al., 2016). Landfills are a major source of atmospheric methane (CH_4), a potent greenhouse gas, contributing significantly to national greenhouse gas emissions. While capturing landfill gas for energy use is a potential solution, it is only cost-effective when there is enough gas (Hettiaratchi et al., 2015).

Landfill bioreactors, a new waste disposal concept, enhance waste decomposition and gas production by recirculating leachate (Reinhart & Townsend, 1998). They operate aerobically or anaerobically. The biocell approach, developed by University of Calgary researchers, combines these modes sequentially. The operation of the biocell as an anaerobic bioreactor enhances CH_4 production, while the aerobic phase enhances waste decomposition to a point when waste can be excavated and mined in a third stage for resource and space recovery, thus making the landfill operation sustainable. This method does not only aid in resource recovery and space utilization but also tackles methane emissions, groundwater

DOI: 10.4324/9781003534129-17

contamination, and energy recovery, potentially creating "zero methane landfills" (Hettiaratchi, 2007). The Calgary Biocell project, implemented in Calgary, Alberta, from 2004 to 2022, aimed to demonstrate the biocell technology in a cold climate, using a pilot-scale waste cell with automated and manual monitoring systems to gather data.

The primary aim of this chapter is to discuss the data and information collected over the 18-year period of the project and to educate the public about sustainable waste management using the biocell approach. This chapter will therefore provide guidance for future biocell research and assist policymakers and investors to assess the biocell concept's role in advancing sustainable waste practices in Canada.

This chapter begins with a general description of the project, followed by the details of the biocell's design, construction, and operation. The following sections focus on landfill biogas production and waste settlement during the biocell's anaerobic phase. The chapter then discusses the final phase of operation, including cell excavation and resource recovery, and presents insights from the analysis of excavated waste to assess its biological characteristics and energy potential. The final section concludes with a summary of the findings.

The Calgary Biocell

Overview

While the concept of landfill bioreactors is well established within the research community, full-scale implementations are still relatively scarce (Reinhart & Townsend, 1998). Only a few full-scale landfill bioreactors, both anaerobic and aerobic, are active globally (see Chowdhury et al., 2024 for a comprehensive list). In North America, some are used as pilot projects to study waste decomposition with leachate recirculation, but they are operated as either aerobic or anaerobic bioreactors. The award-winning Calgary Biocell was notable for integrating anaerobic and aerobic bioreactor processes with landfill mining in a single facility (Hunte et al., 2012). Unlike traditional bioreactor landfills, the Calgary Biocell, was not intended to function as a permanent facility. After complete biodegradation, the contents were excavated and removed, clearing the space for a new biocell.

Despite its potential as an alternative to conventional landfilling, the biocell approach faces technical challenges that may limit its widespread adoption. The Calgary Biocell project sought to address critical questions such as the effects of ambient temperatures on waste degradation, gas production and gas collection. It also focused on the impact of waste settlement on cell space utilization, leachate recirculation and cell infrastructure integrity, maximizing energy and resource recovery, assessing and mitigating methane emissions during cell construction, and evaluating the economic and environmental viability of using excavated waste for energy and landfill biocover. The following sections detail the biocell's design, construction, and operation, highlighting its key features and research capabilities.

Calgary Biocell design features

The Calgary Biocell, located at the City of Calgary's Shepard waste management facility, covered 100 × 100 m with an 85 × 85 m waste footprint and reached 18 m in height with the final cover. It had a pyramidal square frustum shape (see Figure 12.1) with a 3H:1V side slope and consisted of three waste lifts, each separated by a 500-mm-thick cover layer.

Leachate collection and recirculation

The biocell's bottom liner was a composite system with an 80-mil high-density polyethylene (HDPE) geomembrane and a 1.0-m-thick clay liner, protected by a geotextile cushion. The leachate collection system featured pipes spaced 20 m apart, drainage gravel across the cell base, and a geocomposite on the slopes. A 0.6 m deep gravel trench directed leachate to a double-lined sump with a leak detection system and an automated submersible pump for removal. The biocell's liquid injection system included horizontal trenches with perforated pipes in three layers and four vertical HDPE wells extending 3 m above the leachate collection system. The system recirculated leachate and added moisture to maintain optimal waste moisture content.

Landfill gas collection and use

The gas collection system featured horizontal piping and vertical wells. Horizontal pipes were installed in three layers; two beneath intermediate covers and one

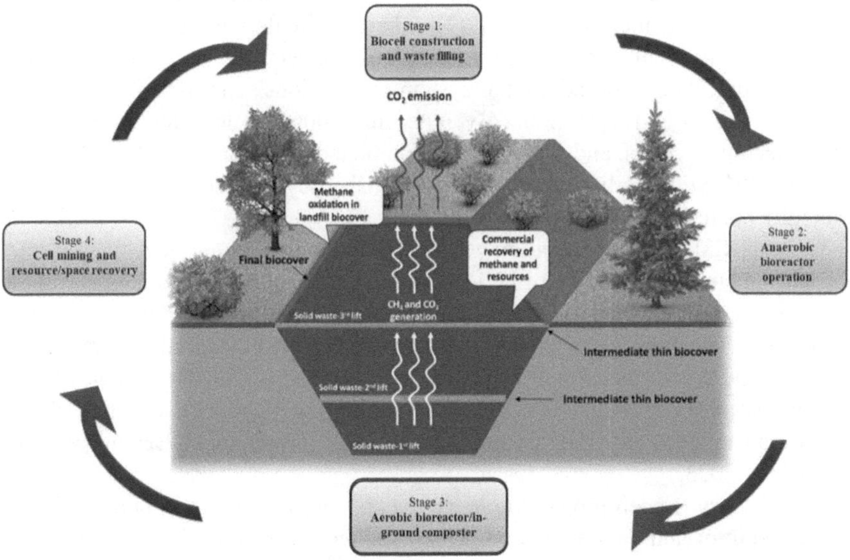

Figure 12.1 Stages of biocell operation.

under the biocap surrounded by coarse aggregate. Four vertical HDPE wells, with perforations at the bottom and solid tops, provided redundancy. Collected gas was directed to the Enmax power grid for power generation, along with gas from the Shepard Landfill, using a 380 kW reciprocating engine.

Biocell construction

Construction of the biocell began in summer 2004. After installing the bottom liner and leachate collection system, it received about 45,000 tons of high-organic residential and commercial waste, placed in three lifts of 5–6 m each. Low-ground-pressure equipment spread the waste, with increasing density from additional layers. Stormwater was sprayed to adjust moisture content. Sensors and load cells were installed, and an initial 1.5 m waste layer was added to protect the infrastructure.

Two intermediate covers were placed on the first and second lifts to separate waste from the environment and act as fire breaks. Made from compost, soil, and mulch, these covers also facilitated methane oxidation and prevented leachate movement. Unlike conventional landfills, the biocell used a hydromulch daily cover instead of low permeable soil. The final cover included a 500-mm-thick compost-based biocover, a geocomposite gas collection layer, a 40 mil LLDPE (linear low-density polyethylene) barrier membrane, a 350 mm subsoil layer, and a 150 mm topsoil layer, designed to enhance gas collection and minimize methane emissions.

Biocell operation and data collection

Waste deposition for the first lift occurred from April to July 2005, followed by the installation of intermediate cover, leachate recirculation, and gas extraction systems. Filling of the second lift resumed in August 2005 and was completed by December 2005, with additional system installations. The third lift was filled between January and April 2006. By June–July 2006, the leachate recirculation, gas collection systems, and sensors were completed. The final cover was added in August 2006, with facility instrumentation finished by October 2006. The biocell was fully operational by November 2006.

Originally, the biocell operation was planned to include three stages: anaerobic, aerobic, and mining (see Figure 12.1). However, a decision was made to eliminate the aerobic stage but the anaerobic operation continued until 2019 to maximize biogas recovery. Because of the rapid degradation of the biodegradable fraction of the waste, by 2019, gas extraction had dropped from a peak of 104 m^3/h in 2007 to approximately 0.2 m^3/h, a 99.8% reduction, prompting plans for excavation. Waste excavation and recovery of excavated waste residue (EWR) took place in Fall 2022 and Winter 2023.

The Calgary Biocell project implemented an extensive monitoring program for process control and data collection. This included tracking on biogas collection, leachate redistribution, waste settlement, and biochemical reactions as indicated by temperature within the cell. Data collection started with waste deposition in June

2005 and continued throughout the filling period as well as postclosure until 2022. Key instruments and their functions included:

- *Thermal mass flow meter*: Measured biogas flow from the biocell.
- *Online gas analyzer*: Monitored methane and oxygen levels using an oxygen sensor and methane detector, with alarms for high oxygen or low methane and connected to an automatic shutoff valve.
- *Butterfly valves*: Estimated biogas flow in branch pipes.
- *Differential pressure gauges*: Measured pressure in pipelines and at butterfly valves.
- *Type T thermocouples and thermistors*: Monitored temperature and moisture.
- *TDR (time domain reflectometry) moisture sensors:* Estimated leachate redistribution; piezometers measured leachate head in the gravel drainage layer.
- *Roctest soil settlement gauges*: Measured settlement with a pressure transducer and antifreeze-filled tubing connected to a reference station.
- *Total pressure cells:* Estimated waste weight at specific points for settlement analysis.
- *LEL (lower explosive limit) meters:* Detected methane leaks with infrared sensors and 1% LEL accuracy (500 ppm CH_4). Two H_2S sensors with 0.25 ppm accuracy detected hydrogen sulfide inside the building.

Biogas production and extraction at the Calgary Biocell

Biogas production and escape during cell filling stage

Although biogas production was suspected immediately after waste placement in the biocell, extraction was not attempted due to logistical issues. Instead, a program measured fugitive gas emissions across intermediate covers by identifying hotspots with surface methane concentration (SMC) measurements, followed by flux chamber measurements.

A surface scan between July 20 and July 25, 2005, detected CH_4 in 37% of 400 locations, with 11% having concentrations between 0.35% and 5%, and 2% exceeding 5%. Fifty-two hotspots with SMCs above 0.35% were selected for flux measurements, showing high CH_4 production initially, with average flux rates of 96±111 g/m²·d and 356±359 g/m²·d for CO_2 (Table 12.1). These values indicate rapid aerobic and anaerobic biodegradation. With typical CH_4 flux rates around 150 g/m²·d in completed landfills, the excessively high values observed at hotspots indicate that preclosure biogas emissions should not be ignored.

Methane fluxes from the second intermediate cover varied widely, with tree mulch areas showing high rates of 147±42 g/m²·d, while a stable compost and tree mulch mixture had much lower rates of 7.9±5.1 g/m²·d. In one area, flux rates ranged from 17.5±41.7 to 65.9±144 g/m²·d across 16 cells over a month. These variations were due to heterogeneity in waste characteristics and the types of intermediate cover used.

Table 12.1 Methane and carbon dioxide flux rates (g/m²d)

Locations	*Date of measurement*	CH_4	CO_2	*# of sample points*
1st intermediate biocover				
	20–25 July 2005	96.3±110.6	356.0±359.4	52
	8 Aug 2005	27.7±36.6	283.9±336.0	16
	12 Aug 2005	41.2±72.1	175.6±185.1	16
	15 Aug 2005	17.5±41.7	171.3±163.1	16
	18 Aug 2005	47.6±91.9	159.9±243.7	16
	19 Aug 2005	65.9±144	182.3±293	16
2nd intermediate biocover				
	23 Dec 2005	7.9±5.1	33.2±10.9	16
	04 Jan 2006	4.8±12.3	161.7±67.5	16
	05 Jan 2006	3.1±4.3	225.6±77.9	16
	25 Jan 2006	4.6±7.2	77.4±47.9	16
	04 Jan 2006	65.2±48.9	388.6±146.5	4
	05 Jan 2006	147.2±41.8	682.1±103.0	4

Source: This table is adapted from Rajbhandari et al. (2006).

Research indicated that biogas production in the biocell can be significant within weeks of filling (Rajbhandari et al., 2006). Compost-based intermediate covers facilitated some biological methane oxidation, reducing CH_4 emissions, while tree mulch covers had little impact. Considering there was no gas collection at this point in time, most biogas produced during cell filling escaped into the atmosphere.

Biogas production and extraction during the first 1,000 days of biocell operation

By August 20, 2009, the operators had collected 1,485,410 m³ of biogas over 1,000 days, averaging a flow rate of 59 m³/h. The average hourly biogas extraction rate varied monthly, reaching a peak of 104 m³/h in June 2007 and a low of 14 m³/h in February 2008. The highest was observed during summer operations and the lowest was observed during winter operations indicating the cold temperatures may have impacted the operation at the biocell. However, a closer analysis indicated that other factors contributed to this variability as well. The key factors that contributed to the variability in the amounts of biogas collected by the operators included the following:

- Leachate management: Automated leachate recirculation improved gas production, whereas manual control led to fluctuations, as seen in August 2007 and February 2008.
- Ambient conditions: Cold temperatures led to pipe freezing, reducing gas collection in Winter 2007 and early 2008. Gas production peaked in Summer.
- Biocell operations: Gas extraction rates varied with vacuum pressure. High pressures typically increased gas recovery, but excessive pressures and mechanical issues like air leaks, reduced production.

The issues with frozen condensate in the manifold lines were resolved by insulating the manifolds to prevent freezing. Gas treatment at the biocell involved a two-stage condensing unit to remove silicates and sulfur compounds, and knock out units for moisture removal.

The collected data (Hunte, 2010), along with known waste composition information, were used to determine kinetic parameters k (decay rate constant) and L_0 (ultimate methane potential) of the biocell. A gas recovery rate of 90% is considered to be reasonable for the biocell considering that it has an impervious geomembrane cover and an active extraction system. The kinetic parameters for the biocell can therefore be estimated as $k = 0.076\ yr^{-1}$ and $L_0 = 103\ m^3$/ton of waste.

In comparison, conventional landfills in Canada generally exhibit much lower decay rates. Alberta Environment (2008) suggested a k value of 0.023 yr^{-1} for Alberta landfills based on expected annual precipitation levels. Although this recommended k value has not been field-verified and might be an overestimate, it remains only 30% of the value determined for the Calgary Biocell. The k value for the biocell is currently the only field-verified kinetic parameter for any landfill type in Alberta.

Temperature fluctuations within the biocell

In Calgary's severe winters, cold temperatures were expected to impact gas production at the biocell. Temperature measurements at various levels within the biocell indicated that while gas temperatures ranged from 3 to 12°C in winter and spring, they rose to about 20°C in summer. Despite ambient temperature fluctuations, the biocell's internal temperatures remained stable, suggesting effective insulation by the waste. The temperatures at the biocell's base showed minimal ambient impact below 5 m from the surface (refer to Hunte, 2010 for temperature profile). Leachate temperatures did not significantly drop, indicating that the waste effectively insulated against cold conditions. Overall, stable temperatures reflect consistent biological activity without immediate overheating risks.

These observations contrast with those of Alghazali et al. (2024), who conducted a 12-year study at a conventional landfill site in Quebec, Canada. They saw a significant negative of ambient temperature on waste degradation and temperature rise, noting a 12–18-month delay in temperature increase during cell filling stages due to frozen or partially frozen waste. Their dataset included three waste lifts placed during freezing conditions, with subsequent lifts set under warmer temperatures. In comparison, biocell waste was placed during summer and fall, so the waste was not frozen during or after placement. The operational differences between the Quebec landfill and the biocell account for the varying observations at the two sites.

Settlement behavior of Calgary Biocell

Landfill bioreactors, with leachate recirculation to increase moisture and microbial biomass within a cell, decompose waste more quickly than conventional landfills, resulting in faster waste settlement. Understanding settlement patterns is crucial for managing cell volume and designing systems for leachate recirculation and gas recovery. The biocell's data on pre- and postclosure waste settlement is unique due to its cold climate context and was used to calibrate and verify a specialized waste settlement prediction model.

Preclosure waste settlement

Settlement occurs both during waste cell construction and cell operation over several years due to waste decomposition, but preclosure settlement is often overlooked. In bioreactor landfills, which use leachate recirculation, this early settlement is crucial. At the biocell, settlement gauges were installed at various points to monitor settlement throughout filling. The gauges used pressure transducers and a reference station to measure changes relative to a known elevation, as shown in Figure 12.2.

Despite three of the sensors being damaged during waste filling, the sensor above the first intermediate cover operated successfully from August 2005 to

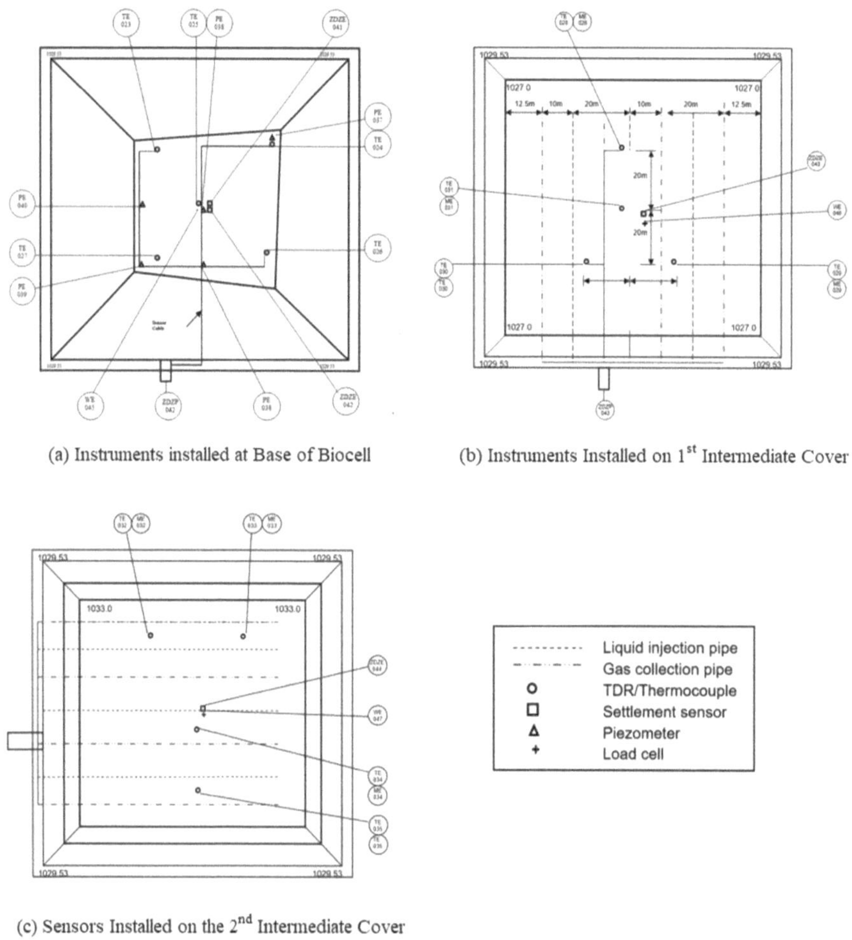

Figure 12.2 Locations of settlement monitoring instruments within the biocell.

Source: This figure is adapted from Hunte (2010).

March 2006, providing valuable data on preclosure settlement. During the filling of the second and third lifts, the first lift settled by about 700 mm, or about 14% strain.

Postclosure waste settlement

Postclosure settlement was tracked using surface profiling with a total station. A grid with 81 markers, spaced 10 m apart, covered an 85 × 85 m area. Surface elevations were monitored from 2006 until the cell excavation in 2022. Figure 12.3(a, b) shows the surface elevations 2 months and nearly 3 years after

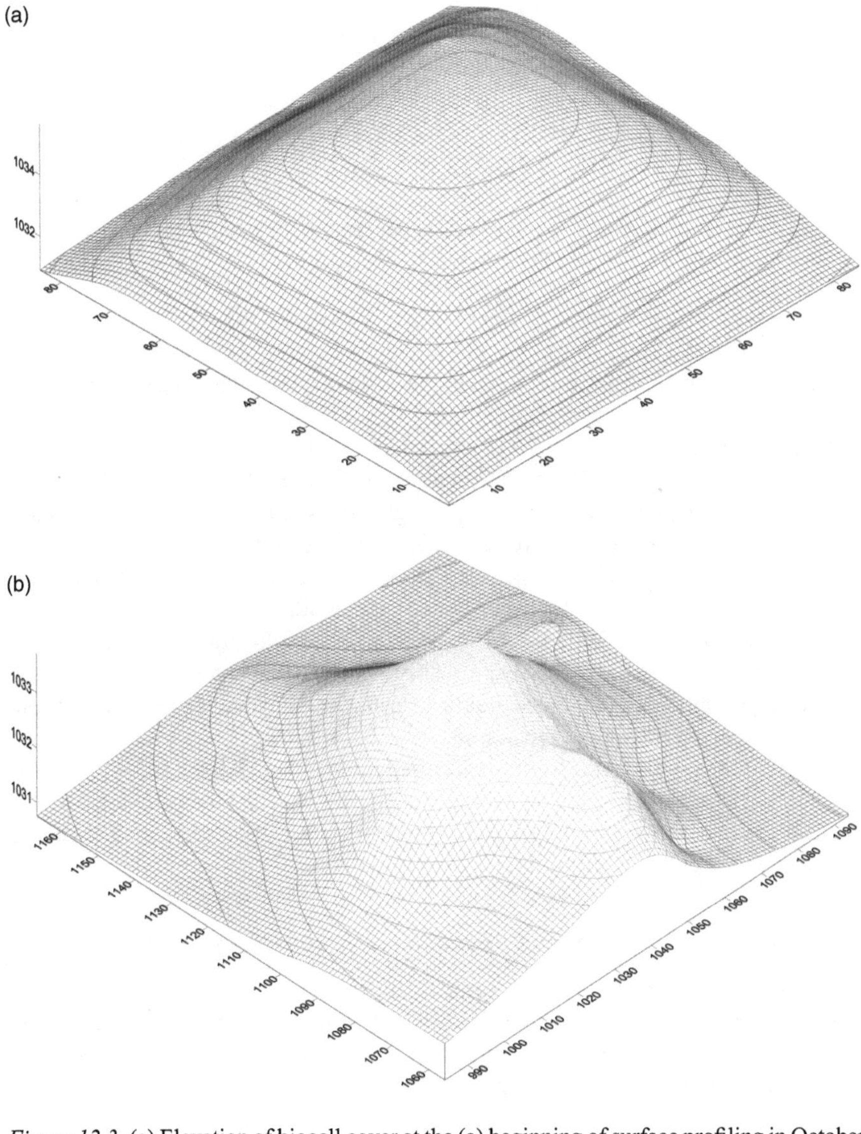

Figure 12.3 (a) Elevation of biocell cover at the (a) beginning of surface profiling in October 2006 and at the (b) end of monitoring period in July 2009.

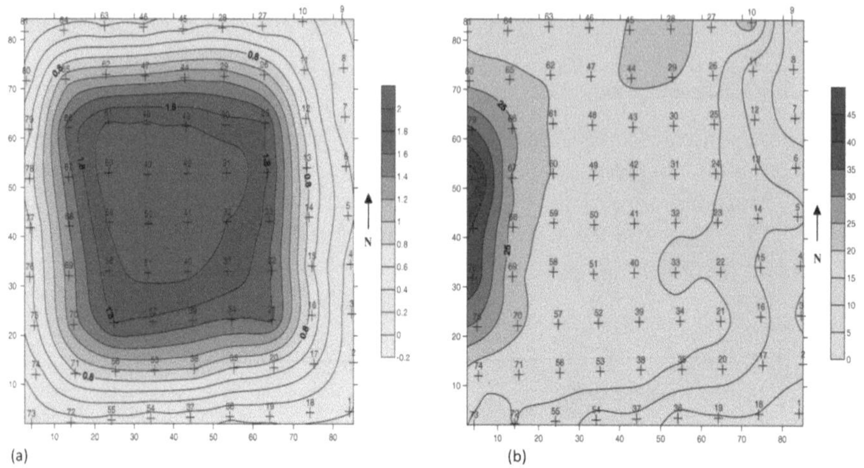

Figure 12.4 (a) Biocell settlements three years after closure (July 2009). (b) Settlement as a percent strain after three years (July 2009).

Source: Adapted from Hettiarachchi et al. (2013).

capping, respectively. Figure 12.4 depict the surface settlement from October 2006 to July 2009, including strain distribution.

Survey data collected from 2007 until 2022 established surface profiles along four cross-sections: A–A, B–B, C–C, and D–D (Figure 12.5). Figure 12.6 shows the A–A cross-section profile, revealing the greatest settlement in the cell's center, where waste was deepest. The western side experienced more settlement than the eastern side due to the cell's deeper base slope and increased moisture from water introduced via sprinklers, which accelerated biodegradation.

Settlement data from 2006 to 2022 were converted into strains to assess settlement accurately. Measurements from grid points #23 and #59, which exhibited higher settlements, were analyzed. Table 12.2 shows the calculated strains at these points and their average over the period.

As of 2022, strain values from monitoring points GP #23, #41, and #59 ranged from 19.82% to 23.48%, with an average of 21.27%. Results indicate rapid settlement in the first five years, followed by gradual stabilization, reflecting dynamic behavior and eventual equilibrium.

A novel mathematical model to predict settlement behavior of bioreactor landfills

Most landfill settlement models neglect enhanced gas generation and moisture distribution effects. The Calgary Biocell project aimed to develop a new model to predict settlement by categorizing waste based on degradation rates. It incorporates mechanical compression and biodegradation, using a linear

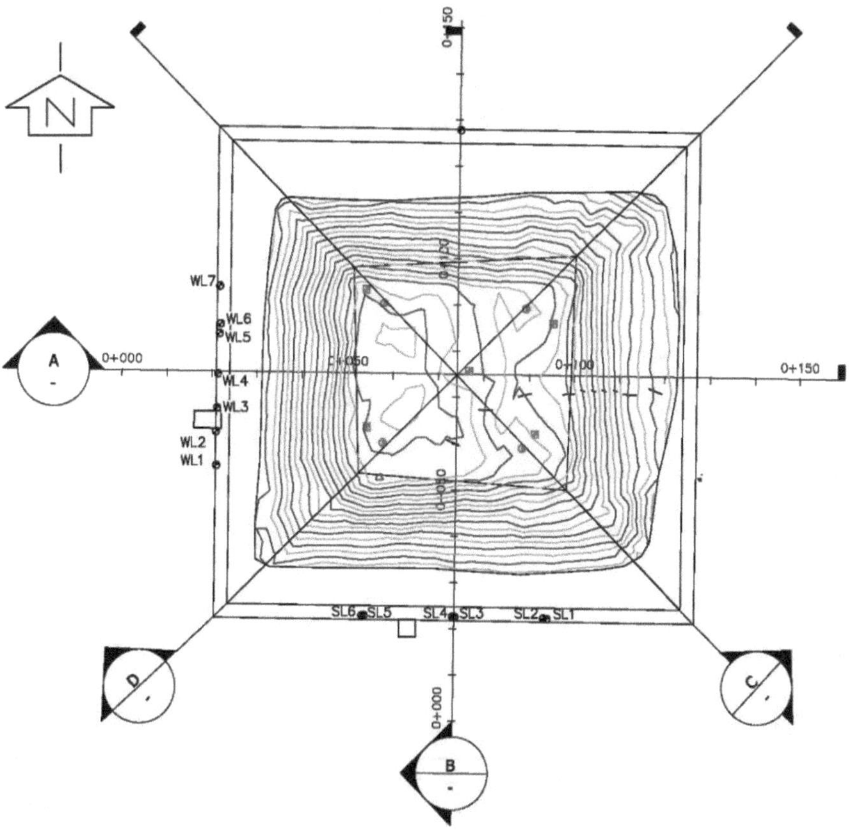

Figure 12.5 Layout of biocell surface profiles.

Source: Adapted from Abeysiri Munasinghage (2024).

stress–strain relationship and first-order kinetics. It also estimates gas pressure and moisture content, adjusting overburden pressure accordingly. Mechanical compression from added waste layers increases stress, while biodegradation decreases mass and stress. Hettiarachchi's model, based on similar principles, also accounts for volume changes due to these factors (Hettiarachchi et al., 2005; Hettiarachchi et al., 2009).

Validated with 16 years of biocell data (2006–22), the model predicted a total strain of 20.18%, 5% lower than the measured 21.27% (Abeysiri Munasinghage, 2024; Munasinghage et al., 2024). Although initial predictions were lower due to waste heterogeneity, accuracy improved over time. Despite consistently underestimating strain, the model effectively captured settlement trends, with errors decreasing as the simulation progressed (see Table 12.3).

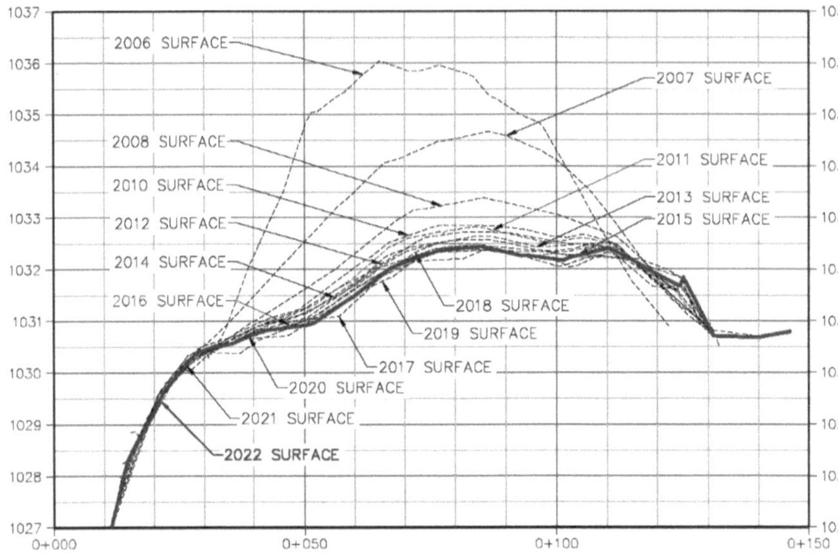

Figure 12.6 Surface profile along cross-section A–A (all dimensions are in meters).

Source: Adapted from Abeysiri Munasinghage (2024).

Table 12.2 Calculated strains at the center of the biocell

Date	*Strain (%)*			
	GP #23	*GP #41*	*GP #59*	*Average*
12-Oct-2006	0	0	0	0.00
25-Apr-2007	5.29	8.34	7.81	7.15
25-Jul-2009	12.23	13.68	14.21	13.37
24-Jun-2010	16.96	18.34	22.74	19.35
12-Apr-2011	17.02	18.57	20.73	18.77
4-Apr-2012	16.49	18.66	21.15	18.77
1-May-2013	18.77	19.57	21.95	20.10
5-May-2014	19.12	19.88	20.82	19.94
6-Aug-2015	18.73	19.96	22.47	20.39
1-Nov-2016	19.83	20.72	22.58	21.04
28-Dec-2017	20.32	20.83	22.41	21.19
5-Nov-2018	19.76	20.49	22.25	20.83
22-Nov-2019	19.88	20.74	22.45	21.02
24-Nov-2020	19.83	22.62	23.45	21.97
9-Oct-2021	19.78	20.49	23.56	21.28
21-Apr-2022	19.82	20.5	23.48	21.27

Source: This table is adapted from Abeysiri Munasinghage (2024).

Table 12.3 Comparison of the measured and model-predicted strain values

Time (days)	*Predicted strain (%)*	*Actual average strain (%)*	*Error (%)*
62	5.06	7.15	29.24
257	9.46	-	-
623	12.55	13.37	6.16
1,079	14.08	19.35	27.24
1,413	15.08	18.77	19.65
1,705	16.04	18.77	14.54
2,063	16.85	20.10	16.18
2,455	17.45	19.94	12.48
2,824	18.07	20.39	11.38
3,282	18.58	21.04	11.73
3,735	18.99	21.19	10.39
4,157	19.26	20.83	7.55
4,469	19.57	21.02	6.93
4,851	19.84	21.97	9.69
5,219	20.06	21.28	5.74
5,538	20.18	21.27	5.10

Source: This table is adapted from Abeysiri Munasinghage (2024).

LBC mining and valorization of EWR

EWR samples and biocell mining

Prior to full-scale mining of the biocell, samples of EWR were gathered in 2020 by borehole drilling using the grab sampling method. Three pilot boreholes were drilled at three different points on the biocell surface. The spatial location of each borehole drilling point was selected based on the distance from the edge of the biocell to determine the effect of spatial change on degradation.

The physical and chemical characteristics of the EWR grab samples are summarized in Table 12.4. The sample numbers refer to three surface areal points and two or three lifts along the biocell depth.

A volatile solids (VS) fraction below 25% indicated the stability of solid waste, suggesting significant degradation in the deeper biocell sections, though some biodegradable material remained (de Araújo Neto et al., 2021). The average lignin content of 17.1% was unlikely to significantly affect waste degradation. Enhanced buffering capacity in the biocell, as indicated by alkalinity, VFA, and pH, supports a stable degradation process. The ultimate analysis estimated the molecular formula for the biodegradable VS as $C_{207}H_{314}O_{141}N$.

Biochemical methane potential of EWR

Key biological properties of EWR, such as biochemical methane potential (BMP) and methane oxidation potential (MOP), are critical for assessing its environmental

Table 12.4 Physical and chemical characteristics of the EWR grab samples

Parameter	*Unit*	*Sample number*							
		$P_{1\text{-}1}$	$P_{1\text{-}2}$	$P_{2\text{-}1}$	$P_{2\text{-}2}$	$P_{2\text{-}3}$	$P_{3\text{-}1}$	$P_{3\text{-}2}$	$P_{3\text{-}3}$
Dry solids (DS)	%TS	58.39	54.70	56.31	55.18	56.07	59.91	55.13	51.39
Volatile solids (VS)	%DS	70.38	54.05	84.41	47.16	60.97	89.64	84.31	39.95
pH	-	7.3	7.5	6.5	6.6	7.4	7.3	6.8	7.2
Lignin	%DS	14.28	21.27	18.01	15.24	16.07	18.54	17.63	16.26
Total phosphorus	g/kg	1.65	3.14	1.79	2.86	0.58	1.91	1.87	2.16
Alkalinity ($CaCO_3$)	g/L	2.32	2.54	2.12	1.98	2.26	2.08	1.94	1.82
Volatile fatty acids	g/L	1.57	1.72	1.93	1.95	1.59	1.01	1.46	1.34
Ammonia-nitrogen (NH_3-N)	g/L	0.005	0.006	0.006	0.011	0.009	0.003	0.005	0.002

Source: This table is adapted from Abedi Yarandy (2022).

and economic value. The BMP was evaluated using a novel method to measure its remaining methane generation potential, which is essential for assessing the biocell's sustainability (Pearse et al., 2018). BMP assays were conducted using a composite sample of EWR, tested in triplicate (CS-1, CS-2, CS-3) under anaerobic, mesophilic conditions with an inoculum from the Bonnybrook WWTP to enhance biodegradation. Twelve BMP assays were prepared: eight for individual EWR samples, three for the composite sample to check variability, and one blank with inoculum only to measure CH_4 generation from the inoculum. Biogas production was measured using water displacement and gas chromatography to determine CH_4 volume.

The BMP assay results show that biocell technology effectively reduces the residual methane potential of landfill waste by enhancing degradation. This helps mitigate greenhouse gas emissions and improves landfill management by promoting thorough organic degradation and resource recovery. Additionally, residual CH_4 production decreased with depth, indicating extensive degradation in deeper layers. The high residual CH_4 potential of samples with high paper and cardboard highlights the need for improved segregation and recycling before biocell processing.

The composite samples had an average BMP_{100} of 34.45 mL CH_4/g VS, lower than conventional landfill waste. This suggests that the biocell's advanced degradation processes effectively reduce CH_4 potential of residual waste excavated from the biocell. Residual CH_4 yields varied by depth, with upper layer samples showing higher potential (39.1 mL CH_4/g VS) compared to lower layers (27.06 mL CH_4/g VS), due to more extensive degradation in deeper layers.

Table 12.5 presents the modified Gompertz model parameters and *k* values for cumulative CH_4 yields in the BMP assays of EWR composite samples. The average residual methane generation potential (L_0) of the composite EWR samples was 16.81 mL CH_4 /g Total Solids (TS). The biocell achieved similar degradation levels in 15 years as a sanitary landfill would in 40 years. Grab samples showed decreasing L_0 values with depth, further supporting the observation that deeper waste had undergone more extensive degradation. Samples with higher paper and cardboard content exhibited lower L_0 values, indicating that these materials contributed more to biogas generation in the biocell.

Table 12.5 Model parameters for EWR composite sample replicates

	P (mL CH_4/ g VS)	*R_m (mL CH_4/g VS/day)*	*λ (day)*	*L_o (mL CH_4/ g VS)*	*k (day^{-1})*
CS-1	41.54	0.63	38.30	15.41	0.0201
CS-2	46.27	0.59	38.02	14.87	0.0197
CS-3	54.33	0.81	38.66	20.15	0.0189

Source: This table is adapted from Abedi Yarandy (2022).

The *k* values were obtained from BMP_{100} data, which was derived from the time-dependent cumulative CH_4 production after 100 days. The average *k* value was 0.0196 day^{-1} and lower than those reported by Fei et al. (2016) and Sandoval-Cobo et al. (2020). Additionally, the lag phase (λ) values were also longer compared to other studies. The longer lag phase may be attributed to the use of the solid-phase method in the BMP assay experiment.

MOP of EWR

The MOP evaluation of the biocell's EWR, using Pokhrel's procedure and Michaelis-Menten kinetics, revealed that while EWR can oxidize methane, it is less efficient than traditional biocover materials (Pokhrel, 2006). This reduced efficiency is partly due to suboptimal nutrient levels and moisture content. Despite these limitations, EWR offers environmental benefits by reducing the need for fresh materials.

The average maximum oxidation rate (V_{max}) for composite samples was 0.43 μmol/h/g (see Table 12.6), lower than traditional materials. Samples from deeper EWR layers, with better moisture conditions, had higher V_{max} values (average 0.76 μmol/h/g) compared to shallower layers (average 0.47 μmol/h/g). Overall, EWR's efficiency was about 30% compared to traditional materials, though deeper samples achieving around 50% efficiency. Supplementing EWR may be needed to enhance methane oxidation performance. While EWR is less efficient than traditional landfill biocovers, it supports environmental sustainability by reducing the need for new materials.

Energy from EWR

While a biocell's space can be recovered through waste residue mining, effective management of the excavated materials is crucial for achieving true sustainability. Steam gasification is a notable treatment technology for this purpose, as it converts the residue into combustible hydrogen rich syngas, which can be utilized as a clean energy resource for various applications. The use of steam as a gasifying agent also results in products with minimal environmental impacts, preventing the formation of NO_x and producing low CO_2 emissions in the gasifier and downstream processes.

Table 12.6 Average V_{max} and K_m of the EWR composite sample

Sample	*Average methane oxidation parameters*	
	V_{max} *(μmol/g/h)*	K_m *(ppm)*
CS-1	0.42	260.52
CS-2	0.40	212.98
CS-3	0.47	232.83

Source: This table is adapted from Abedi Yarandy (2022).

Steam gasification of EWR recorded a maximum CH_4 yield of 11.70 mol/kg, with a lower heating value of syngas reaching 13.38 KJ/Nm^3. Carbon gasification efficiency and cold gas efficiency were 10.64% and 11.38%, respectively. It is worth noting that the LHV (Lower Heating Value) of the EWR in this study is smaller than that of typical excavated residues from conventional sanitary landfills because the biocell is a comparatively more advanced biodegradation facility (Hettiaratchi et al., 2015), leaving less chemical energy in the final excavated residues. This finding is also consistent with the lower fixed carbon content of EWR compared to that of excavated residues from conventional landfills since the fixed carbon content is positively related to the LHV of fuel.

Overall, the results show that EWR has limited potential for energy generation through steam gasification due to its low combustible organic content. However, it could be utilized in a cogasification plant with a higher LHV fuel.

Conclusions

The biocell concept represents an innovative advancement in sustainable waste management, offering significant improvements over traditional landfilling methods. Its applicability in northern climates was demonstrated by successful completion of the Calgary Biocell project. Operational from 2006 to 2022, the Calgary Biocell underwent extensive performance studies, generating valuable data. Over the last two years, the stabilized waste within the biocell was excavated, and the evaluation of the EWR revealed its potential use as landfill biocap materials and shows promise for energy generation via steam gasification. Key findings from the Calgary Biocell project include:

- **Waste degradation and biogas production**: The biocell's leachate recirculation and biogas collection systems greatly enhanced waste degradation, biogas production, resource recovery, and emissions control. However, severe cold temperatures can affect biogas collection unless proper precautions are taken. Cold weather has minimal impact on microbial waste degradation kinetics.
- **Real-time data monitoring**: Advanced real-time monitoring of temperature, moisture, settlement, and leachate offered valuable insights into the biocell's operations, though cold weather can affect sensor durability.
- **EWR valorization**: Analyses of the biocell's EWR showed better biodegradation than conventional landfills, with potential uses in landfill biocover material and cogasification for energy recovery.
- **Sustainable waste management**: The biocell model demonstrated significant potential for sustainable waste management through improved resource recovery, reduced greenhouse gas emissions, and the reclamation of waste cell areas for future use.

Overall, the biocell approach provides a transformative alternative to traditional waste management by integrating waste reduction, reuse, and recovery, potentially revolutionizing the field.

References

Abedi Yarandy, T. (2022). *Biochemical methane potential and methane oxidation potential of excavated waste residue from Calgary Biocell* (Master's thesis, University of Calgary, Calgary). https://prism.ucalgary.ca/items/0b0eb5e9-84a4-4a8a-9125-8e35666da15d.

Abeysiri Munasinghage, S. A. (2024). *Settlement behavior of bioreactor landfills* (Master's thesis, University of Calgary, Calgary). https://prism.ucalgary.ca/items/7a18ecb7-a351-42ca-954e-40201bcfee67.

Alberta Environment. (2008). *Technical guidance for the quantification of specified gas emissions from landfills*. https://open.alberta.ca/dataset/7894c055-e4c3-426e-a254-0a54da1beef0/resource/9b775382-ce70-494d-8c9a-166d1846fcc3/download/2008-11-technicalguidancedocument-landfill.pdf

Alghazali, W., Kaur, S., Van Geel, P. J., & Kenny, S. (2024). A comprehensive study of temperature data during the filling and post-closure phases at a landfill in Quebec, Canada: Application of a thermal-mechanical-biological model. *Waste Management & Research*, 1234–1245. https://doi.org/10.1177/0734242X241270938.

Chowdhury, S. D., Mishra, A., Surampalli, R. Y., Zhang, T. C., Kao, J. C. M., Bhunia, P., & Hettiaratchi, J. P. A. (2024). A critical review on bioreactor landfills: Types, design, leachate recirculation, leachate and landfill gas quality, reaction kinetics, current implementation scenario, sustainability, challenges, and future perspectives. *Journal of Cleaner Production* (in review).

de Araújo Neto, C. L., Gurjão, R. Í. L., & Farias, A. P. (2021). Physical, chemical, and mechanical properties of landfilled waste from Campina Grande, Brazil. *Environmental Science and Pollution Research*, 28(63), 66819–66829. https://doi.org/10.1007/s11356-021-15230-5.

Fei, X., Zekkos, D., & Raskin, L. (2016). Quantification of parameters influencing methane generation due to biodegradation of municipal solid waste in landfills and laboratory experiments. *Waste Management*, 55, 276–287. https://doi.org/10.1016/j.wasman.2015.10.015.

Gao, J., Oloibiri, V., Chys, M., Audenaert, W., Decostere, B., He, Y., Van Langenhove, H., Demeestere, K., & Van Hulle, S. W. (2015). The present status of landfill leachate treatment and its development trend from a technological point of view. *Reviews in Environmental Science and BioTechnology*, 14(1), 93–122. https://doi.org/10.1007/s11157-014-9349-z

Hettiarachchi, H., Meegoda, J. N., & Hettiaratchi, J. P. A. (2005). Towards a fundamental model to predict the settlements in bioreactor landfills. GeoFrontiers 2005, Waste Containment and Remediation, ASCE Publishing, 1–15. https://doi.org/10.1061/40789(168)51.

Hettiarachchi, H., Meegoda, J., & Hettiaratchi, P. (2009). Effects of gas and moisture on modeling of bioreactor landfill settlement. *Waste Management*, 29(3), 1018–1025. https://doi.org/10.1016/j.wasman.2008.08.018.

Hettiarachchi, H., Hettiaratchi, J. P. A., Hunte, C. A., & Meegoda, J. N. (2013). Operation of a landfill bioreactor in a cold climate: early results and lessons learned. *Journal of Hazardous Toxic and Radioactive Waste*, 17(4), 307–316. https://doi.org/10.1061/(asce)hz.2153-5515.0000159.

Hettiaratchi, J. P. A. (2007, September). New trends in waste management: North American perspective. In *Proceedings of the international conference on sustainable solid waste management* (pp. 9–14).

Hettiaratchi, J. P. A., Jayasinghe, P.A., Tay, J., & Yadav, S. (2015). Recent advances of biomass waste to gas using landfill bioreactor technology: A review. *Current Organic Chemistry*, 19(3), 413–422. https://doi.org/10.2174/1385272819666150119223002.

Hunte, C. A. (2010). *Performance of a full-scale bioreactor landfill* (Doctoral thesis, University of Calgary, Calgary). http://dx.doi.org/10.11575/PRISM/3647.

Hunte, C. A., Hettiarachchi, C. H., Meegoda, J. N., & Hettiaratchi, J. P. A. (2012). The City of Calgary biocell landfill: Data collection and settlement predictions using a multiphase model. In *Geo-congress 2012: State of the art and practice in geotechnical engineering* (pp. 4202–4211). https://doi.org/10.1061/9780784412121.432.

Meegoda, J. N., Hettiarachchi, H., & Hettiaratchi, J. P. (2016). Landfill design and operation. In J. Wong & R. Surampalli (Eds.), *Sustainable solid waste management* (pp. 577–604). American Society of Civil Engineering.

Munasinghage, S., Hettiarachchi, H., Hettiaratchi, P., & Wong, R. C. K. (2024). Settlement analysis of Calgary landfill biocell: An anaerobic landfill bioreactor operated in cold climates for 15 years. In *A&WMA's 117th annual conference & exhibition*, Calgary, Alberta.

Pearse, L. F., Hettiaratchi, J. P., & Kumar, S. (2018). Towards developing a representative biochemical methane potential (BMP) assay for landfilled municipal solid waste: A review. *Bioresource Technology*, 254, 312–324. https://doi.org/10.1016/j.biortech.2018.01.069.

Pokhrel, D. (2006). *Compost based biocap performance* (Doctoral thesis, University of Calgary, Calgary). https://prism.ucalgary.ca.doi:10.11575/PRISM/928.

Rajbhandari, B. K., Hettiaratchi, J. P. A., & Hurtado, O. (2006). Thin biocovers (TBC): A novel approach in controlling methane emissions from landfills accepting biodegradable organic waste. In *Proceedings of the 21st international conference on solid waste technology and management*, Philadelphia, PA.

Reinhart, D. R., & Townsend, T. G. (1998). Full-scale experiences with bioreactor landfills: Case studies. In *Landfill bioreactor design and operation* (1st ed., pp. 61–84). Routledge. https://doi.org/10.1201/9780203749555-4

Sandoval-Cobo, J. J., Casallas-Ojeda, M. R., Carabalí-Orejuela, L., Muñoz-Chávez, A., Caicedo-Concha, D. M., Marmolejo-Rebellón, L. F., & Torres-Lozada, P. (2020). Methane potential and degradation kinetics of fresh and excavated municipal solid waste from a tropical landfill in Colombia. *Sustainable Environment Research*, 30(1), 1–11. https://doi.org/10.1186/s42834-020-00048-6.

Siddique, A., Hahladakis, J. N., & Al-Attiya, W. A. K. (2022). An overview of the environmental pollution and health effects associated with waste landfilling and open dumping. *Environmental Science and Pollution Research*, 29(39), 58514–58536. https://doi.org/10.1007/s11356-022-21578-z.

Zohoori, M., & Ghani, A. (2017). Municipal solid waste management challenges and problems for cities in low-income and developing countries. *International Journal of Science and Engineering Applications*, 6(2), 39–48. https://doi.org/10.7753/IJSEA0602.1002.

13 Phosphorus recovery from food waste

A case study of Canadian solid waste management practices

Pradeep Thapa, Shailja Pant, Prangya Ranjan Rout and Qiuyan Yuan

Introduction

Food is a fundamental necessity for human existence. A typical life cycle of food starts with the growth of a plant and includes its harvesting, handling, storage, processing, distribution and consumption (Gooch et al., 2019). However, with the global population projected to reach 9.8 billion by 2050 (United Nations Department of Economic and Social Affairs, 2017), the current rate of production and supply of food will be inadequate for future demand. As a result, there is an ever-increasing demand of higher food production, which consequently leads to increased use of mineral fertilizers. Among the various mineral fertilizers utilized to increase production and yield of plant, phosphorus is one of the most essential fertilizers that significantly affects plant growth (Cordell et al., 2011; Papangelou et al., 2020). Global phosphorus fertilizer use has risen significantly, from 4.6 million tons (MT) in 1961 to approximately 21 MT in 2015, with future estimates suggesting that global demand could reach 22–27 MT per year for cropland and an additional 4–12 MT for grassland by 2050 (Bindraban et al., 2020). However, phosphorus reserves may deplete within next 350 years, with peak demand projected around 2030 (Walsh et al., 2023).

Food loss occurs wholly or partially at various stages such as harvesting, handling, storage, distribution and consumption, leading to generation of waste that ultimately becomes inedible (Cordell et al., 2011; Gooch et al., 2019). Interestingly, this food waste contains at least one-fifth of the phosphorus used in food production by the time it reaches consumers, with even larger quantities lost in earlier stages (Cordell et al., 2009). In lieu of all these facts, recently, attention has been shifted toward repurposing of food waste from being an invaluable waste stream to a valuable source of phosphorus. Implementation of recycling and recovery methods can lead to effective phosphorus recovery, thus mitigating the challenge of phosphorus scarcity while contributing to a more sustainable agricultural approach (Cordell et al., 2011). The major sources of phosphorus for animals include their diets consisting of other animals and plants while plants derive phosphorus from soil. Over the years, the continued crop production without adequate phosphorus replenishment has led to depletion of phosphorus levels in soil, resulting in

DOI: 10.4324/9781003534129-18

decreased food productivity and potential famines (Cordell & White, 2014; Suh & Yee, 2011). This scenario created a need for many countries in Europe and Asia to look for local alternatives such as animal manure, bones, crop residues, human excreta, etc. to boost the soil phosphorus level and substantially increase plant growth and productivity (Cordell et al., 2009; Cordell & White, 2014). The current flow of phosphorus in the global food system is demonstrated in Figure 13.1.

Globally, approximately 12 MT of phosphorus is harvested annually in the form of crops; of which, around 7 MT phosphorus is processed into feed, food and fiber, while approximately 5 MT is returned to the land in the form of crop residues (Cordell et al., 2009) . Furthermore, 50% of the phosphorus consumed by livestock is returned to agriculture land through excretion (Cordell & White, 2014). However, the data do not represent a comprehensive global picture because of the massive disparity in phosphorus use between developed and developing countries. Although the demand for phosphorus from developing countries (especially Asia) is two-third of the global demand, the weak purchasing power of these countries limits access to the phosphorous resources (Food and Agricultural Organization, 2015). In Africa, more than 75% of agricultural soil is phosphorus-deficient, consequently resulting in decreased crop yield (Cordell et al., 2009) whereas developed regions such as Europe and North America experience phosphorus overuse through overfertilization of arable land because of the purchasing power of farmers and readily available high-grade phosphorus (Cordell & White, 2014). This deficiency

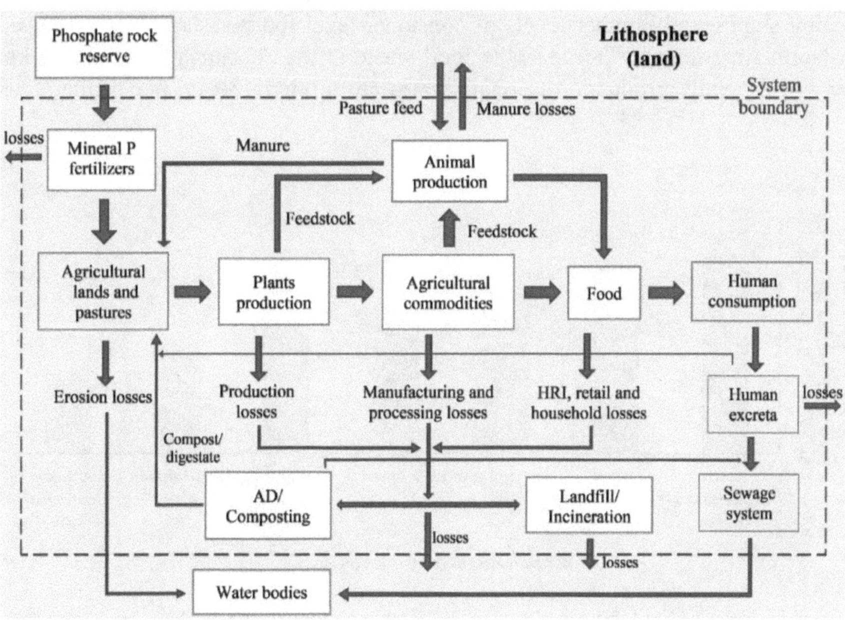

Figure 13.1 Global phosphorus flow in food production and consumption.

highlights a critical imbalance in phosphorus availability and its consumption; making recovery efforts an urgent imperative toward restoration of the imbalance.

Although North America is typically characterized by phosphorus overuse, inefficiencies in food production and high levels of food waste still result in significant phosphorus losses (Jacob et al., 2017); Canada, in particular, faces substantial food waste issues (Gooch et al., 2019) making it an ideal context for studying phosphorus recovery from food waste. This chapter focuses on phosphorus recovery in Canada and evaluates the best methods for maximizing the theoretical recovery potential from food waste. This is crucial given the country's considerable food waste losses. By recovering phosphorus from food waste, Canada can help mitigate the environmental impacts of phosphorus overuse while improving the sustainability of its agricultural systems.

Food loss and waste across the world

A study conducted by the Food and Agricultural Organization in 2011 concluded that one-third of the food produced globally—worth $200 billion by weight and approximately 1.2–2 billion tons—is wasted each year (Food and Agriculture Organization, 2011; Paul et al., 2014). This loss is primarily because of the differences in food production and consumption patterns globally. Figure 13.2 illustrates the per capita food wastage reported by the Food and Agriculture Organization (2011) across different regions of the world; broken down into two categories: wastage from production to retail and wastage at the consumer level. High-income regions such as North America, Oceania, Europe, Industrialized Asia exhibit significantly higher levels of consumer-level food wastage. For instance, in North America and Oceania, the food waste at the consumer level is 115 kg per capita, while production-to-retail losses are around 180 kg per capita. The

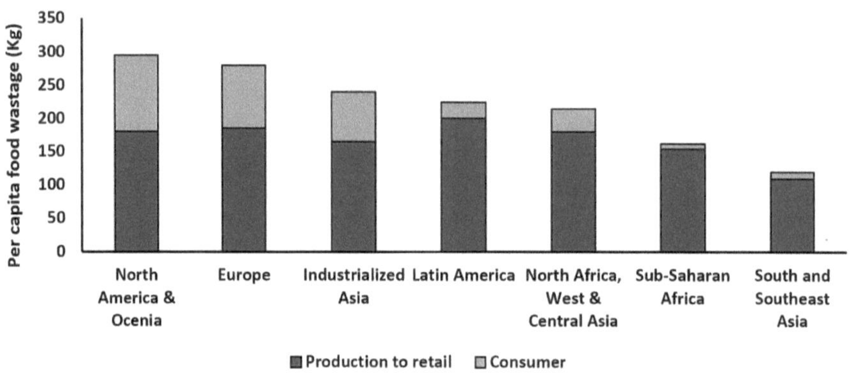

Figure 13.2 Per capita food wastage across different regions of the world.

Source: Food and Agricultural Organization (2011).

same trend is observed in Europe and Industrialized Asia, though at slightly lower per capita values (95 and 75 kg, respectively). In contrast, developing regions including Latin America, North Africa, sub-Saharan Africa, South and Southeast Asia experience lower consumer-level waste. For example, sub-Saharan Africa records only 10 kg per capita of consumer food waste, compared to 155 kg per capita lost between production and retail. In regions like South and Southeast Asia, consumer wastage is similarly low at 10 kg per capita, but losses from production to retail are significant at 110 kg per capita. Such a scenario is a consequence of the confusion among consumers over the "best before" and "expiry" dates (Gooch et al., 2019). According to the report *Global food losses and food waste* by Food and Agriculture Organization (2011) another possible reason for the uneven pattern of food waste or losses for different parts of the world is the implementation of inferior harvesting technologies, transportation, inadequate storage facilities, etc., that take place during various stages of the food supply chain.

Poor harvesting techniques that lead to damaged crops degrade food quality, making the crops unsellable, which in turn contributes to the generation of food waste. This wasted food, despite containing valuable nutrients like phosphorus, ends up in landfills or is discarded before reaching consumers. Similarly, poor infrastructure for transportation and distribution can result in food contamination by pests and other vectors, further increasing the amount of food waste (Gooch et al., 2019). These inefficiencies do not only result in the loss of food but also in the significant loss of phosphorus embedded in these products. Recovering phosphorus from such waste streams becomes crucial, as this nutrient can be reused to support future agricultural productivity and reduce the strain on nonrenewable phosphorus reserves.

Need for food waste treatment/phosphorus recycling

The global distribution of food waste underscores the need for region-specific phosphorus recovery strategies. In high-income regions, efforts should focus on improving consumer behavior and implementing food waste treatment systems to recover lost phosphorus. In developing regions, optimizing agricultural practices and infrastructure for phosphorus recovery at earlier stages of the food supply chain could provide a crucial boost to soil fertility and food security. By addressing food waste through a phosphorus recovery lens, both developed and developing regions can contribute to more sustainable food systems while reducing the strain on global phosphorus reserves.

The availability of phosphorus is concentrated in a limited number of countries such as Morocco, China, Russia, and the USA, which together control more than 75% of the total reserve phosphate ore (Campos et al., 2019). Alarmingly, up to 55% of phosphorus is lost from "farm to fork" (Cordell, 2009), due to inefficiencies like overproduction, improper food handling, and premature harvesting (FAO, 2011). These losses do not only represent wasted resources but also contribute to environmental degradation. With advanced waste treatment technologies such as anaerobic digestion (AD), these sources can be effectively processed to extract

Table 13.1 Sources of phosphorus for recovery

Organic matter	*P (% P by weight)*	*Source of P*
Activated sewage sludge	1.4	Human waste (excreta)
Crop residue	0.04–0.33	Agricultural waste
Farmyard manure	0.07–0.88	Agricultural waste
Human excreta	0.35	Human waste
Poultry manure	1.27	Agricultural waste
Sludge (from biogas digester)	0.48–0.77	Agricultural &(or) food &(or) human waste
Vermicompost	0.65	Food waste

Source: Cordell et al. (2011).

phosphorus and reintroduce it into the agricultural system (Bolzonella et al., 2018). Table 13.1 highlights the amount of recoverable phosphorus from multiple sources, underscoring the significance of integrating phosphorus recovery into food management systems.

In 2016, North American countries generated 2.21 kg of waste per capita per day, amounting to 289 MT annually (Kaza et al., 2018). Food waste is a significant component of organic waste that accounts for 28% of this total. As food waste degrades, it not only loses its inherent value—up to 2.5 times its original worth—but also requires considerable resources such as energy, water, land, and infrastructure for disposal (Gooch & Felfel, 2014). These operations come with high costs, with labor constituting a major fraction of food waste management expenses (WRAP, 2018). Moreover, food waste disposal in landfills produces greenhouse gases (GHGs), which contribute to climate change. When factoring in GHG emissions and the loss of nutrients like nitrogen (N), phosphorus (P), and potassium (K), global food waste losses are estimated to exceed $1 trillion annually and could rise to $2.6 trillion by 2030 (Slorach et al., 2020).

Efficient food waste treatment methods—such as composting and anaerobic digestion—offer solutions to recover phosphorus and help meet the growing global demand for this nutrient (Cordell et al., 2011; Papargyropoulou et al., 2014). These methods do not only recover phosphorus but also generate biogas and energy, contributing to environmental sustainability (Vandermeersch et al., 2014). Given the increasing demand for phosphorus in global food production, both short- and long-term management strategies are needed to ensure phosphorus is reused and recovered effectively. However, current regulations in many regions, including Canada, do not mandate optimal phosphorus recovery (Cordell et al., 2011).

In addition, rising phosphorus use has led to widespread nutrient pollution country in both land and water systems in Canada, exacerbating environmental degradation (Cordell, 2014). Excessive phosphorus runoff from agricultural lands has been a significant issue, particularly in regions like the Great Lakes, where it has led to harmful algal blooms and degraded water quality (Schindler et al., 2012). This highlights the urgent need for more sustainable phosphorus management and recovery practices in Canada through advancements in phosphorus recovery technologies and implementing stricter mandates to reduce both nutrient loss and environmental impact.

Food waste and phosphorus recovery scenario in Canada

Canada was ranked the eighth worst among 36 countries within the Organization for Economic Co-operation and Development (OECD) index in 2019 based on the effectiveness of waste management per capita (Solid Waste & Recycling, 2019). In 2002, Canada directed 24.1 MT of solid waste to the landfills (Environment and Climate Change Canada, 2023). Fast forward to 2022, the amount of waste sent to landfills was increased by 8% to reach 26.1 MT. Over this same period, the total solid waste generated grew by 5.3 MT (17%) to reach 36.0 MT (Environment and Climate Change Canada, 2023). On a positive note, the amount of waste diverted for recycling improved significantly, increasing by 3.3 MT (49%) to reach 9.9 MT by 2020 (Environment and Climate Change Canada, 2023). However, despite these improvements, a large portion of organic waste, including food waste, continues to end up in landfills. Given that organic waste constituted approximately 30% of the total solid waste generated in Canada in 2020 (Statistics Canada, 2020), there is a critical need to improve diversion rates, especially for food waste.

Though there was a significant improvement in the reduction of waste per capita, the amount of waste diverted for recycling was not encouraging. In 2019, the amount of waste diverted from landfills was just about 145 kg of waste per capita, that is, 20% of the total waste (Solid Waste & Recycling, 2019). According to the report *Taking stock: Reducing food loss and waste in Canada* published by Environment and Climate Change Canada in 2019, organic and kitchen waste constitutes approximately 30% of the total solid waste disposed by households in Canada.

Food waste represents a more significant portion of this organics waste. In Canada, as much as 51% of the food waste is generated at home, which is equivalent to 40% of the total food produced and worth as much as $27 billion (Paul et al., 2014). Additionally, food waste occurs along various stages of a food distribution system (Gooch et al., 2019). This amount is expected to increase in the future. Therefore, food waste diversion from landfills is of paramount importance today. According to Gooch et al. (2019), the action required for waste management is to identify, reduce and reuse nonreusable and nonedible food waste. Furthermore, the food waste diverted from landfills should be treated to retrieve useful resources such as biogas, nutrients (Nitrogen, Phosphorus and Potassium) etc. In order to manage these diverted organics, two main treatments, that is, AD and composting, are being widely used all over the world. Composting has long been established as a primary method of organic waste treatment in Canada, with widespread adoption in both backyard and large-scale municipal programs. However, AD is increasingly emerging as a complementary technology for the treatment of organic waste, particularly food waste (Canadian Biogas Association, 2018). While composting continues to play a critical role in organic waste management, anaerobic digestion offers additional benefits, such as biogas production and nutrient recovery, such as nitrogen, phosphorus, and potassium). As both technologies advance, they represent complementary solutions in the broader framework of waste diversion and resource recovery.

In Canada, composting remains a more popular method of organic waste treatment than AD. By 2020, approximately 63% of Canadian households participated in some form of composting, including source-separated organics (SSO) programs for kitchen waste and yard waste (Statistics Canada, 2021). For instance, Prince Edward Island (PEI) led the way with 95% of kitchen waste being composted, followed closely by Nova Scotia at 94% (Vanderkloet, 2023). Despite its larger population, Ontario reported only 62% of kitchen waste being composted (Statistics Canada, 2021). Composting has several advantages over landfilling, including lower operational costs, reduced greenhouse gas emissions, and the production of valuable compost that enriches soil (Paul et al., 2014). Studies have shown that backyard composting, in particular, has the lowest operational costs compared to landfilling, making it both an environmentally and economically viable option (Blair et al., 2014). In contrast, AD is emerging as a complementary technology, particularly for food waste treatment. According to the report published by Canadian Biogas Association, *Biogas and renewable natural gas in Ontario: 2019 market overview and outlook*, as of 2019 there were over 60 operational AD facilities in Canada, with Ontario housing 42 of these facilities. These AD plants primarily processed livestock manure (52%) and food waste (33%), while some used sewage sludge as feedstock (Canadian Biogas Association, 2019). AD facilities play a significant role in biogas production, with agricultural sectors contributing 68% of total biogas, which is used to generate 810 MW of electricity annually (Canadian Biogas Association, 2019).

Moreover, AD reduces greenhouse gas emissions and produces digestate, a nutrient-rich byproduct that can be applied to farmlands, reducing the need for chemical fertilizers (Slorach et al., 2020). The Canadian government, along with provincial initiatives, has supported the development of both composting and AD. For instance, Nova Scotia, Ontario, and PEI are leading provinces in adopting organics diversion programs where policies such as landfill bans for organic waste and public awareness campaigns are being formulated to expand these programs (Ministry of Environment and Climate Change, 2018). Additionally, partnerships between the Canadian Food and Inspection Agency (CFIA) and the National Zero Waste Council (NZWC) are promoting research on energy recovery and material reuse from organic waste streams (National Zero Waste Council, 2016). These initiatives demonstrate the growing focus on sustainable food waste treatment in Canada, which can reduce landfill dependency, recover valuable nutrients like phosphorus, and generate renewable energy through biogas production.

Case study: System definition

A significant portion of the food wastes at various levels of food production and supply chain in Canada suggests potential recovery through multiple treatment

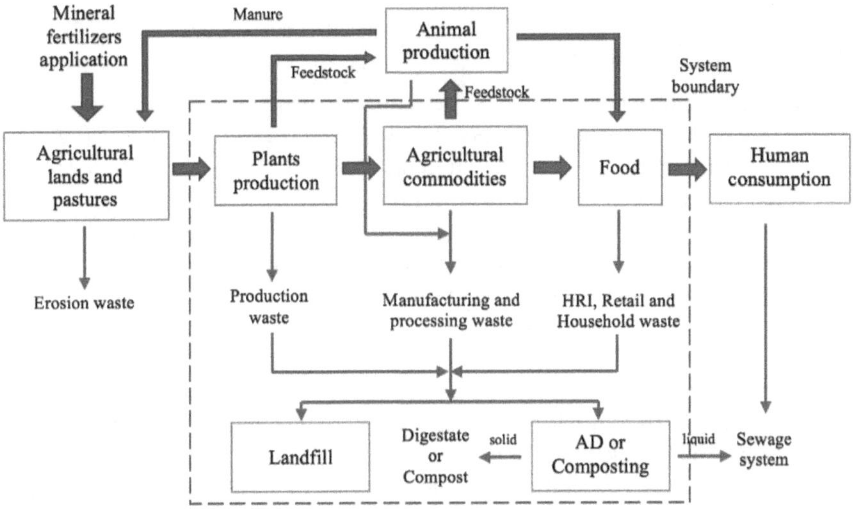

Figure 13.3 System diagram of phosphorus flow in Canada.

methods. This study takes the flow of phosphorus through the food supply chain in Canada into consideration. The system boundary in this analysis is defined by the general food supply chain logic. Here, the flow of phosphorus from plants to humans and ultimately to their end life is considered. The complete flow is illustrated in Figure 13.3. The system boundary focuses on production, trade, and consumption stages, where food production and consumption patterns are globally consistent. These stages allow for quantification of potentially recoverable phosphorus, making the findings applicable to global contexts if adequate data is available. In the production stage, losses occur during plant growth and harvesting, including residual stalks, roots, leaves, husks, shells, and straw—collectively referred to as crop residue. These residues are not factored into the analysis since they are typically returned to the fields through processes like chopping, spreading, and burning, allowing the phosphorus content to be reabsorbed into the soil (BC, 2014; Cordell et al., 2009). Additionally, data on phosphorus content in crop residues is limited. After harvesting, these agricultural commodities are further processed and manufactured into a new food product. Additionally, meats are also processed and manufactured. During these processes, processing and manufacturing losses are acquired. Similarly, during trading, processed and manufactured food is sold or served at different places such as stores, markets, restaurants, etc. incurring further wastes at the hotel, restaurant, and institutional (HRI) level and household level. Phosphorus flow along these various system components is determined using MFA (material flow analysis). These flows are for Canada only and given a different situation of food waste across different parts of the world would require reimplementation (FAO, 2011). As such, the results obtained might not apply elsewhere.

Quantification of flows

Phosphorous flow calculation

In this study, phosphorus flow is determined using the MFA (Brunner & Rechberger, 2016), which quantifies the flow of a substance and material within a system. First, the phosphorus amount in the food was quantified. With the available amount of phosphorus present in different foods, the amount of individual phosphorus flow through that system was determined. For the calculation, the amount of phosphorus present in the food was multiplied by the total amount of that food supplied (through production, initial stock and imports). This procedure was utilized to determine the total phosphorus flow from one level of the food supply chain to another in Canada for the years 2015 and 2017. The amount of phosphorus present at the beginning of each stage was determined by the phosphorus quantity present in a certain food type published in the literature. For instance, total phosphorus amount during production, import and beginning stocks in 2015 were determined as 23,800, 5,700, and 1,200 tons, and in 2017 were determined as 25,040, 6,410, and 970 tons, respectively. Similarly, food supplied, available and wasted along each stage of the supply chain was also determined. For the determination of waste, the average loss or waste suggested by Gooch et al. (2019) was utilized. Figure 13.3 illustrates the movement of phosphorus from food supply chains to a household setting. Furthermore, the obtained waste was averaged to determine their individual contribution to the recovery stage.

Phosphorus recovery scenarios

Three scenarios were developed to determine the optimum food waste treatment method for maximum theoretical recovery of phosphorus. Total household food waste of 2015 (functional unit) was considered as the feedstock (i.e., 2,570 tons) for all three treatment processes to determine the maximum theoretical recovery potential. In Canada, household wastes are collected via curbside collection system, but most of them collect organic and nonorganic waste together. In Canada, household waste is typically collected via curbside systems, where both organic and nonorganic waste are often mixed. For the purposes of this study, however, we assumed that all household waste would be collected in a source-separated form, meaning organics and nonorganics were separated at the source. This assumption is necessary to ensure a consistent and homogenous feedstock, as SSO are easier to process and allow for more accurate phosphorus recovery estimates. Without this assumption, variations in contamination rates (e.g., mixing nonorganic materials with organics) would significantly reduce the efficiency of phosphorus recovery, leading to underestimation of the recovery potential. Additionally, many municipalities in Canada are moving toward mandatory source separation, making this assumption relevant for future policy considerations. Hence, the feedstock is assumed to be homogenous and consistent, and that the obtained phosphorus recovery value is accurate. Another assumption is that the accumulated food waste was transported to their respective treatment sites through trucks. Although transportation and handling costs are not included in this analysis, this exclusion is justified by the study's focus on assessing theoretical phosphorus recovery potential. Transportation costs are highly variable and depend on multiple factors, including fuel prices, geographic distances, and municipal policies. However, these costs could be incorporated in future studies to provide a full economic life cycle analysis, centered on technical recovery potential.

Scenario 1

This scenario is indicative of the current situation in Canada, where more than 90% of the collected household food waste is landfilled. Here, most of the food waste goes to the landfill along with the incorporated phosphorus. The rest of the waste is assumed to be diverted to recovery facilities, that is, mostly composting.

Scenario 2

In this scenario, 95% of household food waste is assumed to be diverted to AD facilities. The remaining 5% of the food waste goes to the landfill. This AD process includes a pretreatment of waste as per Bernstad and la Cour Jansen (2011), where food waste bags are opened mechanically, and the organics are further processed, and the rejects (inorganics, plastics, etc.) are landfilled. Thus, obtained food waste fraction is then treated under the mesophilic condition to produce digestate. The finally retrieved byproduct of AD, that is, digestate's phosphorus value, is shown in Table 13.2.

Scenario 3

This scenario considers 95% of food waste being directly composted. The remaining 5% of household food waste is assumed to be landfilled, like in scenario 2. This scenario considers 95% of food waste being directly composted, with the remaining 5% assumed to be landfilled, similar to scenario 2. The composting process is divided into two subcategories. In the first, food waste is collected in a source-separated form and treated in a decentralized composting facility. A structural material of 200 kg of woodchip per ton of waste is added, as per Bernstad and la Cour Jansen (2011). In the second process, a 4:1 ratio of food waste to wood chip/leaves is maintained to ensure stable aerobic composting, as described by Keng et al. (2020). The phosphorus content in the woodchip was not included in the phosphorus recovery calculation. Additionally, the cost of using woodchip as a structural material, including its procurement, transport, and addition to the composting process, was not included in the cost estimate. This exclusion is due to the focus of the study on theoretical phosphorus recovery rather than a full cost analysis.

Phosphorus available in food waste for recovery

Phosphorus flow in Canada demonstrated a linear pattern. In 2015, a total of 25.69 MT of food was injected in the Canadian food system through production, import and previous stocks (beginning stocks), which had a phosphorus footprint of 30,700 tons. Similarly, in 2017, 26.63 MT of incoming food had a phosphorus footprint of 32,420 tons; a high proportion of this phosphorus was lost to exports, that is, 34% in 2015 and 36% in 2017. Food wasted at various stages (processing, manufacturing, retail, HRI and household) equaled 9.28 and 9.63 MT in 2015 and 2017, respectively. Furthermore, the amount of phosphorus available through these sources was 11,800 and 12,580 tons in 2015 and 2017, respectively. The highest amount of phosphorus was available through the processing waste, which equaled to almost 40%, and the lowest was through retail, which contributed to about 6% of the total average recoverable phosphorus. On average, household waste contributed to about 22% of the total recoverable phosphorus.

Figures 13.4 and 13.5 illustrate an increase in the food waste pattern in Canada and, moreover, the increasing phosphorus in waste suggesting the need for the recovery of phosphorus being wasted as an urgent imperative. Food waste technologies such as AD and composting can help in recirculating almost 39% of the total infused phosphorus from the Canadian food system (Figures 13.4 and 13.5).

The best scenario for phosphorus recovery

With 90% of food waste being landfilled, scenario 1 had the lowest possible potential of phosphorus recovery, however, it was still able to recover 10% of the phosphorus through composting. All the phosphorus that was landfilled was lost. In scenario 2, the 95% treated food waste through AD was able to recover 81% of the

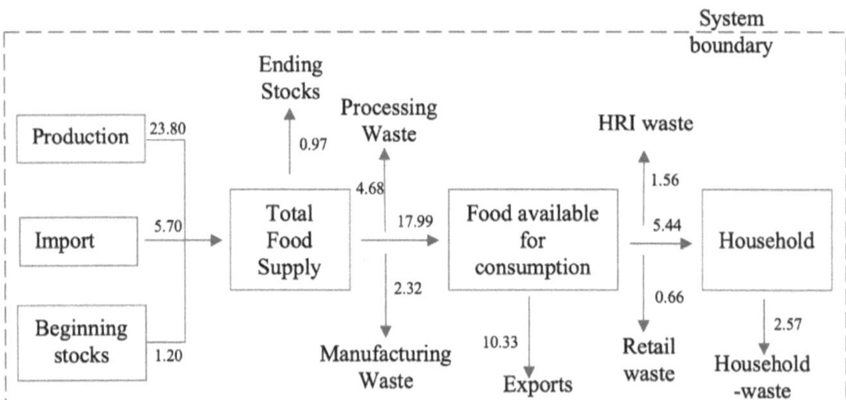

Figure 13.4 Phosphorus flow in 2015 (in thousand tons).

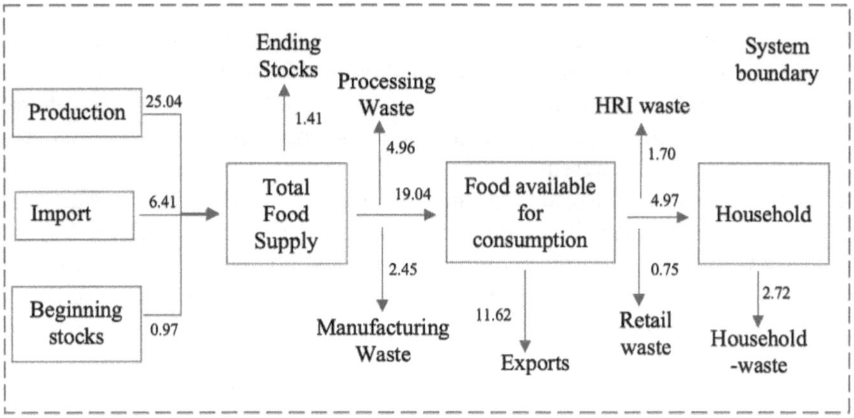

Figure 13.5 Phosphorus flow in 2017 (in thousand tons).

Table 13.2 Phosphorus recovery potential

Scenarios	*Source P- content (1,000 tons)*	*Theoretical P recovered (1,000 tons)*	*P recovery (%)*	*Referred scenarios*
Scenario 1	2.57	0.26[a,b]	10.11	[a]Bernstad & la Cour Jansen, 2011; [b]Keng et al., 2020
Scenario 2	2.57	1.97[a]	76.65	[a]Bernstad & la Cour Jansen, 2011
Scenario 3	2.57	2.44[a,b]	94.94	[a]Bernstad & la Cour Jansen, 2011; [b]Keng et al., 2020

total phosphorus content. This partial loss could be because of the partial retention of phosphorus nutrients due to precipitation in the digester. The potential amount of recovery is, however, also dependent on the process and feedstock of AD. A higher recovery potential has been reported by mixing of cattle manure and using sewage sludge (Möller & Müller, 2012). However, the presence of cattle manure increases the solubility of phosphorus and hence makes it readily available for plant uptake (Möller & Müller, 2012).

The highest potential of phosphorus recovery was demonstrated by scenario 3, that is, composting. The obtained result indicated that almost all the phosphorus in the feed was recovered through the digestate. Table 13.2 reflects the amount of theoretical phosphorus that can be recovered using scenarios 1, 2 and 3.

Conclusion

This chapter examined the potential of phosphorus recovery from food waste in Canada and evaluated the best possible method of treating food waste to acquire the maximum theoretical value of phosphorus recovery. The amount of phosphorus present in various food waste by category, such as processing, retail waste, etc. has been determined in the present study by calculating the phosphorus content in individual food products. The result indicated the potential of recovering more than 7% of the phosphorus utilized in the agricultural sectors in Canada through food waste. The calculation considered the phosphorus content of household food waste for the years 2015 and 2017 to determine the best method of food waste treatment among three present food waste scenarios. Among them, scenario 3, that is, composting, was able to recover 100% of phosphorus present in food waste in the form of compost. Similarly, scenario 2, that is, AD, was able to recover 81% of the phosphorus from food waste in the form of digestate. The obtained compost and digestate can be used in agricultural fields to replace synthetic phosphorus fertilizers. The least capable food treatment scenario was landfilling, scenario 1, which accounted for 90% of phosphorus loss. The results provide an insight into the phosphorus recovery potential in Canada. Public awareness and behavior change will be essential in encouraging greater adoption of composting and AD. Future studies can be directed toward more detailed analysis and research to fully understand the scalability and implementation of these recovery methods on a national scale.

References

BC. (2014). *Strengthening farming: Right to farm*. Ministry of Agriculture, British Columbia. www2.gov.bc.ca/assets/gov/farming-natural-resources-and-industry/agriculture-and-seafood/agricultural-land-and-environment/strengthening-farming/farm-practices/870218-30_crop_residue_management.pdf. Retrieved 16-07-2020.

Bernstad, A., & la Cour Jansen, J. (2011). A life cycle approach to the management of household food waste: A Swedish full-scale case study. *Waste Management*, 31(8), 1879–1896.

Bindraban, P. S., Dimkpa, C. O. & Pandey, R. (2020). Exploring phosphorus fertilizers and fertilization strategies for improved human and environmental health. *Biology and Fertility of Soils*, 56(3), 299–317. https://doi.org/10.1007/s00374-019-01430-2.

Blair, A., Hollands, G., McIntosh, K., MacDonald, A., Mehta, B., Umali, H., & Pagsuyoin, S. (2014, April). Alternative management of organic waste in Chatham-Kent, Ontario, Canada. In *2014 systems and information engineering design symposium* (pp. 74–77). IEEE. https://doi.org/10.1109/SIEDS.2014.6829905.

Bolzonella, D., Fatone, F., Gottardo, M., & Frison, N. (2018). Nutrients recovery from anaerobic digestate of agro-waste: Techno-economic assessment of fullscale applications. *Journal of Environmental Management*, 216, 111–119. https://doi.org/10.1016/j.jenvman.2017.08.026.

Brunner, P. H., & Rechberger, H. (2016). *Handbook of Material Flow Analysis: For Environmental, Resource, and Waste Engineers* (2nd ed.). CRC Press. https://doi.org/10.1201/9781315313450

Campos Gómez, J. L., Crutchik, D., Franchi, Ó., Pavissich, J. P., Belmonte, M., Pedrouso Fuentes, A., & Val del Río, M. Á. (2019). *Nitrogen and phosphorus recovery from anaerobically pretreated agro-food wastes: A review*. https://doi.org/10.3389/fsufs.2018.00091.

Canadian Biogas Association. (2018). *Current status and future potential of biogas production from Canada's agriculture and agri-food sector*. https://biogasassociation.ca/index.php/resources/canadian_agricultural_biogas_study.

Canadian Biogas Association. (2019). *Biogas and renewable natural gas in Ontario: 2019 market overview and outlook*. www.biogasassociation.ca/images/uploads/documents/2019/2019-Market-Overview.pdf. Retrieved 6-10-2024.

Cordell, D., Drangert, J. O., & White, S. (2009). The story of phosphorus: Global food security and food for thought. *Global Environmental Change*, 19(2), 292–305. https://doi.org/10.1016/j.gloenvcha.2008.10.009.

Cordell, D., Rosemarin, A., Schröder, J. J., & Smit, A. L. (2011). Towards global phosphorus security: A systems framework for phosphorus recovery and reuse options. *Chemosphere*, 84(6), 747–758. https://doi.org/10.1016/j.chemosphere.2011.02.032.

Cordell, D., & White, S. (2014). Life's bottleneck: sustaining the world's phosphorus for a food secure future. *Annual Review of Environment and Resources*, 39(1), 161–188. https://doi.org/10.1146/annurev-environ-010213-113300.

Curry, N., & Pillay, P. (2012). Biogas prediction and design of a food waste to energy system for the urban environment. *Renewable Energy*, 41, 200–209. https://doi.org/10.1016/j.renene.2011.10.019.

Environment and Climate Change Canada. (2023). *Solid Waste Diversion and Disposal*. www.canada.ca/en/environment-climate-change/services/environmental-indicators/solid-waste-diversion-disposal.html. Retrieved 22-02-2025

Food and Agriculture Organization (FAO) (2011). *Global Food Losses and Food Waste – Extent, Causes and Prevention*. Rome, Italy: FAO. www.fao.org/4/mb060e/mb060e00.htm. Retrieved 22-02-2025

Food and Agriculture Organization (FAO). (2015). *World Fertilizer Trends and Outlook to 2018*. Rome, Italy: FAO. https://openknowledge.fao.org/handle/20.500.14283/5202. Retrieved 22-02-2025

Gooch, M., Bucknell, D., LaPlain, D., Dent, B., Whitehead, P., Felfel, A., Nikkel, L., & Maguire, M. (2019). *The avoidable crisis of food waste: Technical report*. Value Chain Management International and Second Harvest. www.SecondHarvest.ca/Research.

Gooch, M., & Felfel, A. (2014). *$27 billion revisited: The cost of Canada's annual food waste*. https://doi.org/10.1002/joc.1982.

Jacobs, B., Cordell, D., Chin, J., & Rowe, H. (2017). Towards phosphorus sustainability in North America: A model for transformational change. *Environmental Science & Policy*, 77, 151–159. https://doi.org/10.1016/j.envsci.2017.08.009.

Kaza, S., Yao, L., Bhada-Tata, P., & Woerden, F. V. (2018). *What a waste 2.0: A global snapshot of solid waste management to 2050*. Urban Development Series. https://datatopics.worldbank.org/what-a-waste/trends_in_solid_waste_management.html.

Keng, Z. X., Chong, S., Ng, C. G., Ridzuan, N. I., Hanson, S., Pan, G. T., … & Lam, H. L. (2020). Community-scale composting for food waste: A life-cycle assessment-supported case study. *Journal of Cleaner Production*, 261, 121220. https://doi.org/10.1016/j.jclepro.2020.121220.

Ministry of Environment and Climate Change. (2018). *Ontario's food and organic waste framework: Action plan*. https://files.ontario.ca/food_and_organic_waste_framework.pdf. Retrieved 24-3-2020.

Möller, K., & Müller, T. (2012). Effects of anaerobic digestion on digestate nutrient availability and crop growth: A review. *Engineering in Life Sciences*, 12(3), 242–257. https://doi.org/10.1002/elsc.201100085

Papangelou, A., Achten, W. M. J., & Mathijs, E. (2020). Phosphorus and energy flows through the food system of Brussels Capital Region. *Resources, Conservation and Recycling*, 156, 104687. https://doi.org/10.1016/j.resconrec.2020.104687.

Papargyropoulou, E., Lozano, R., Steinberger, J. K., Wright, N., & Ujang, Z. (2014). The food waste hierarchy as a framework for the management of food surplus and food waste. *Journal of Cleaner Production*, 76, 106–115. https://doi.org/10.1016/j.jclepro.2014.04.020

Paul, S., Dubey, B., & Dutta, A. (2014). *Evaluation of a hybrid approach of food waste management*. American Society of Agricultural and Biological.

Slorach, P. C., Jeswani, H. K., Cuéllar-Franca, R., & Azapagic, A. (2020). Assessing the economic and environmental sustainability of household food waste management in the UK: Current situation and future scenarios. *Science of the Total Environment*, 710, 135580. https://doi.org/10.1016/j.scitotenv.2019.135580.

Solid Waste & Recycling. (2019). *Canada scores eighth worst in OECD waste index*. www.solidwastemag.com/solid-waste/canada-scores-eighth-worst-in-global-waste-index/1003282901/. Retrieved 12-03-2020.

Statistics Canada. (2020). *Waste management industry survey: Business and government sectors, 2018*. www150.statcan.gc.ca/n1/en/catalogue/16F0023X. Retrieved 22-02-2025

Statistics Canada. (2021). *The dirt on composting*. www.statcan.gc.ca/o1/en/plus/1327-dirt-composting. Retrieved 22-02-2025.

Suh, S., & Yee, S. (2011). Phosphorus use-efficiency of agriculture and food system in the US. *Chemosphere*, 84(6), 806–813. https://doi.org/10.1016/j.chemosphere.2011.01.051.

United Nations, Department of Economic and Social Affairs, Population Division. (2017). *World Population Prospects: The 2017 Revision*. Retrieved from https://desapublications.un.org/file/624/download

Vanderkloet, R. (2023). *Organic waste diversion in Atlantic Canada*. https://doi.org/10.20381/ruor-29078.

Vandermeersch, T., Alvarenga, R. A. F., Ragaert, P., & Dewulf, J. (2014). Environmental sustainability assessment of food waste valorization options. *Resources, Conservation and Recycling*, 87, 57–64. https://doi.org/10.1016/j.resconrec.2014.03.008.

Walsh, M., Schenk, G., & Schmidt, S. (2023). Realising the circular phosphorus economy delivers for sustainable development goals. *NPJ Sustainable Agriculture*, 1(1), 2. https://doi.org/10.1038/s44264-023-00002-0.

WRAP & World Resources Institute. (2018). *The Business Case for Reducing Food Loss and Waste: Restaurants*. https://wrap.org.uk/resources/report/business-case-reducing-food-loss-and-waste-restaurants. Retrieved 22-02-2025.

14 Analyzing geospatial characteristics and solar potential of closed landfill sites in Saskatchewan

Anica Tasnim, Amy Richter and Kelvin Tsun Wai Ng

Introduction

The management and rehabilitation of closed landfill sites has become a significant concern due to increasing urbanization and waste generation (Weng et al., 2015; Nai et al., 2021). As more landfills reach capacity and cease operations, they pose potential environmental risks and land use challenges (Srivastava et al., 2015; Zuberi et al., 2015). However, these sites also present unique opportunities for renewable energy development, particularly solar power. In Canada, there are over 3,000 municipal solid waste landfills, with more than half now closed (Environment and Climate Change Canada, 2024). Transforming landfills into solar farms addresses postclosure management by minimizing maintenance, reducing costs, overcoming land use challenges, and contributing to sustainable energy production, aligning waste management with broader energy goals (Szabó et al., 2017; Jacob & Ayers, 2018).

This chapter examines the geospatial characteristics of closed landfill sites and explores their potential for solar energy generation. By analyzing factors such as location, topography, and solar radiation, it estimates their solar energy potential, providing practical insights for policymakers and developers to optimize site selection and project planning.

Management of closed landfill sites in Canada

Canada employs a multifaceted strategy for managing closed landfills, with a strong focus on environmental protection and methane reduction (Environment and Climate Change Canada, 2022). Additionally, the Government of Canada has outlined a regulatory framework to control landfill methane emissions, which includes installing infrastructure to recover and utilize landfill gas (Environment and Climate Change Canada, 2023). In Canada, landfill regulation is primarily managed at the provincial and territorial levels, with federal jurisdiction delegated accordingly. Each province and territory develops its own waste management policies and oversees landfill operations (Environment and Climate Change Canada, 2024). In Saskatchewan, proactive measures are evident in the management of closed landfills. For example, Saskatoon has advanced its landfill gas collection and

DOI: 10.4324/9781003534129-19

power generation system by integrating solar panels, contributing to the city's goal of sourcing 10% of its energy from local renewable sources. Similarly, municipalities across Canada are adopting innovative strategies to repurpose landfills. In Dawson City, Yukon, an old landfill has been transformed into a solar farm, generating 280 MWh annually. This initiative significantly reduces greenhouse gas emissions and decreases reliance on diesel (Government of Yukon, 2021). Furthermore, innovative solar technologies, such as those developed by Solar FlexRack, are facilitating the installation of photovoltaic systems on landfill surfaces (Solar FlexRack, 2022), thereby enhancing the viability of renewable energy projects. This approach does not only support renewable energy projects but also transforms potential environmental liabilities into valuable assets. Effective public policies are crucial in supporting landfill management projects and redeveloping these sites into parks, residential areas, or solar farms, underscoring the need for ongoing research in this domain.

Topographic and engineering considerations for solar farms at closed landfills

Studies have demonstrated that closed landfills can be detected and characterized using historical topographic maps, aerial photographs, digital elevation models, and lidar techniques (Stohr & Filippini, 2018; Sliusar et al., 2022). Geographic Information System (GIS) is crucial for assessing landfill suitability for ecological rehabilitation and environmental sustainability (Richter et al., 2019). In particular, topographic considerations such as slope, elevation, and aspect are essential for evaluating land for solar farm installations (Ali et al., 2019; Doorga et al., 2019). Slope impacts panel orientation and stability, elevation affects solar exposure and construction logistics, and aspect optimizes sunlight capture, particularly on south-facing slopes in the Northern Hemisphere (Tansel et al., 2013; Quesada et al., 2015; Günen, 2021). Surface conditions like soil stability and composition also influence the feasibility of solar panel installation (Hernandez et al., 2014). For successful solar deployment on landfills, several factors must be addressed: the age of the landfill (affecting settlement rates), usable acreage (determined through aerial maps and site measurements), cap characteristics (depth and components), maintenance requirements, liability concerns, site control (including fencing and safety), solar resource availability, and solar access (ensuring a minimum of six hours of sunlight), as well as proximity to transmission and distribution lines (Doorga et al., 2019; US Environmental Protection Agency, 2023). Analyzing these factors with GIS and digital elevation models (DEMs) facilitates effective site selection and design for solar energy projects.

Online tools for calculating solar radiation

Solar radiation, which includes visible light, ultraviolet light, and infrared radiation, is the electromagnetic energy emitted by the sun (Singh et al., 2020). Accurate calculation of solar radiation is crucial for assessing the solar energy potential of closed landfill sites. The amount of solar radiation received at a specific location directly influences the feasibility and efficiency of solar energy systems (Kabir et al., 2018). Precise measurement and analysis of solar radiation are fundamental to designing and optimizing solar power installations. Various online tools have been developed to aid in calculating solar radiation and assessing solar energy potential. These tools offer user-friendly interfaces for inputting site-specific parameters such as geographical location, time of year, weather conditions, topography, and albedo effect, all of which influence the amount of solar radiation reaching the earth's surface (Department of Energy, 2024; National Renewable Energy Laboratory, 2024). Many researchers have incorporated these tools to validate their models or simply to calculate solar radiation. For instance, Melhem and Shaker (2023) used the PVWatts calculator to determine the optimal tilt angle for solar panels in GCC countries, demonstrating the calculator's effectiveness in optimizing tilt angles to maximize energy output.

The PVWatts calculator provides monthly, seasonal, and yearly irradiance values, with a percentage error of around 3% compared to mathematical models. Similarly, Arab et al. (2020) used the PVWatts model to calculate solar radiation and predict the performance of photovoltaic (PV) modules, noting its practicality for estimating solar radiation and optimizing PV system performance. The Global Solar Atlas, an online tool provided by the World Bank Group, offers an overview of solar energy potential for any location worldwide. Chu and Hawkes (2020) discussed the use of the Global Solar Atlas to simulate hourly Global Horizontal Irradiance, Direct Normal Irradiance, and Diffuse Horizontal Irradiance. They highlighted the importance of accurate input data and consistent methods to improve the spatial resolution of solar radiation assessments for evaluating solar energy technologies. In Canada, Okunlola et al. (2022) used the Global Solar Atlas to collect Global Tilted Irradiation data for assessing solar-based hydrogen production potential. They highlighted the tool's effectiveness in integrating the impact of optimal tilt angles and ground reflection for better yield. Other calculators, such as the Solar Electricity Calculator, PVGIS, PVMAPS, and RETScreen, are also used to determine the optimal angular tilt and orientation of PV modules. These tools calculate optimal PV module settings based on solar irradiation variations across different climate seasons (Huld, 2017; Buzra & Serdari, 2023).

Methodology

Case study: Closed landfills in Saskatchewan

Saskatchewan is recognized for its significant potential to increase solar and wind capacity (Government of Canada, 2022) due to its vast open spaces, which offers an ideal setting for solar energy production (Natural Resources Canada, 2023). According to Natural Resources Canada (2023), Saskatchewan receives more solar irradiation than any other province or territory in Canada, with an average solar system producing 1,330 kWh of electricity per kW of solar panels per year. Saskatchewan indeed has the highest number of landfills per capita in Canada, with over 500 closed landfills (Government of Saskatchewan, 2024).

Managing landfills in their postclosure phase involves addressing issues such as soil contamination, methane emissions, and groundwater pollution (Zahari et al., 2022). However, the conversion to solar farms can be an effective solution to these challenges. Solar installations require minimal ground disturbance, reducing the risk of contaminant release. Moreover, capping the landfill with a solar array can help contain methane emissions, which can then turn a potential hazard into a controlled environment (US Environmental Protection Agency, 2023). Repurposing these sites can contribute to the region's renewable energy targets, supporting Canada's commitment to reducing greenhouse gas emissions and combating climate change (Government of Canada, 2022). The successful implementation of solar farms on closed landfills in Saskatchewan could serve as a model for other regions, promoting environmental stewardship and sustainable energy practices.

Site selection

Saskatchewan, Canada is divided into 18 census divisions for statistical purposes. This study focuses on census divisions 6, 7, 9 and 10, covering an area of 60,738 km^2. Figure 14.1 shows the study area in Saskatchewan. These divisions offer a representative sample of Saskatchewan's diverse geography, encompassing a range of population centers, a balanced mix of rural and urban settings, and land use patterns. The presence of larger urban centers within these divisions drives higher demand for renewable energy, making solar farm projects both viable and impactful. In Saskatchewan, public opinion largely favors solar panel arrays, with minimal opposition, as they are viewed as a positive step toward reducing carbon footprints and generating clean energy (Pesâkâstêw Solar Project, 2024). Data on landfill locations, sourced from the Government of Saskatchewan's GeoHub, reveals that Division 6 and Division 11 each have five closed landfills, Division 10 has seven, and Division 9, the highest, has 23 closed landfills with permits that expired between 2006 and 2022 (Government of Saskatchewan, 2024). A total of 40 landfill sites were evaluated for their topographic suitability for solar farm development.

Methods

To assess the potential of closed landfills for solar farm installations, a straightforward approach combining remote sensing techniques and GIS tools was implemented. Google Earth Pro was utilized to calculate the area of the closed landfills. Landfills with an area exceeding 5 acres were selected for further analysis. As noted by Sukumaran et al. (2022), a minimum of 3 acres is required for a solar farm. Topographical data were sourced from DEMs provided by the United States Geological Survey (USGS) EarthExplorer. ArcGIS was then used to calculate the mean values of key parameters such as slope, aspect, and elevation at each landfill site. The suitability of these parameters for solar farm installation was assessed based on established criteria:

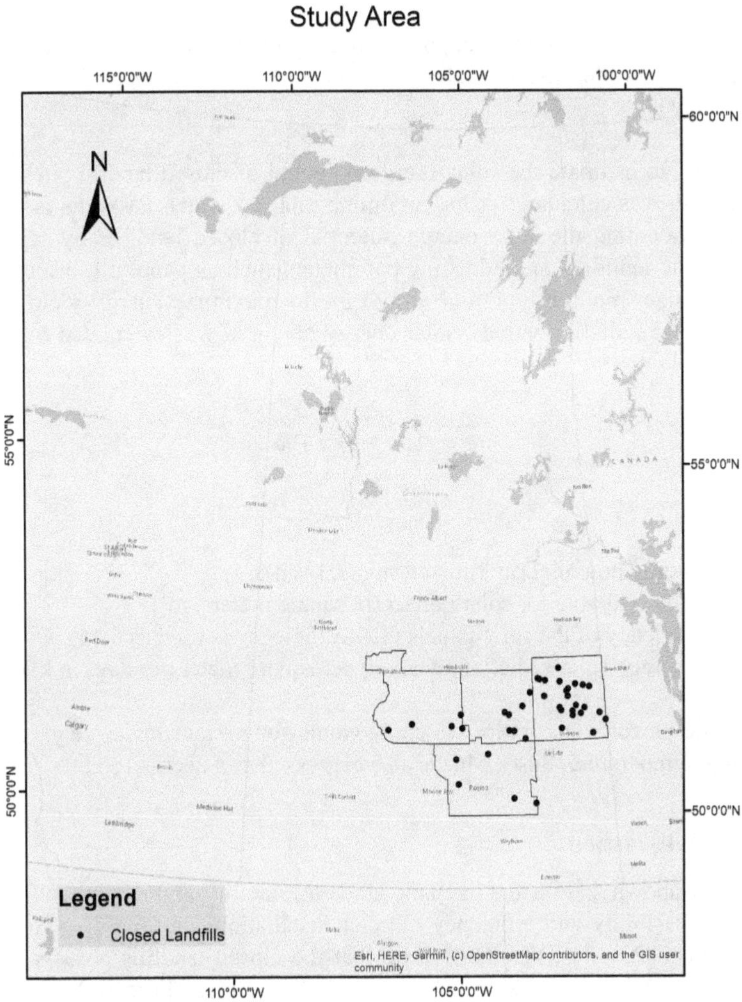

Figure 14.1 Closed landfills site in Divisions 6, 9, 10 and 11 in Saskatchewan.

Source: Data from Government of Saskatchewan (2024).

- Slope: Solar panels can be installed on slopes up to 15% (approximately 8.5 degrees) without significant issues (Rediske et al., 2020; Sukumaran et al., 2022). In this study, a slope of less than 5 degrees is considered ideal for solar farms to ensure maximum sunlight exposure and ease of construction.
- Aspect: In the Northern Hemisphere, south-facing slopes are preferred for solar redevelopment due to their optimal sunlight exposure year round. Williams et al. (2023) identified slopes with an aspect of 135–225 degrees as highly suitable

and those with an aspect of 90–135 degrees as moderately suitable for solar projects.

- Elevation: Elevation should not pose significant challenges for construction and maintenance while ensuring adequate solar radiation (Munsell, 2013; Tansel et al., 2013).

Moreover, to estimate the solar energy potential of closed landfill sites, solar radiation data was calculated using an online tool, PVWatts. PVWatts is instrumental in evaluating the solar energy potential of closed landfills by selecting location of the landfills, and adjusting parameters such as panel tilt, orientation, DC system size, module type and array type to maximize energy yield. After collecting solar radiation values, solar energy potential was calculated by using Equation 14.1:

$$E = A \times r \times H \times PR \tag{14.1}$$

where

- E is the energy potential (in kilowatt-hours, kWh).
- A is the area available for solar panels (in square meters, m^2).
- r is the efficiency of the solar panels (15% efficiency is used as 0.15).
- H is the average solar radiation received per square meter per day (in $kWh/m^2/day$).
- PR is the performance ratio, which accounts for system losses (e.g., due to shading, temperature, dust). This is also expressed as a decimal (0.80).

Results and discussion

Studies have shown that factors like land gradient, stability, and slope significantly impact the feasibility and efficiency of solar installations on landfills (Moodliar & Davidson, 2023). The results indicate that the closed landfills possess favorable topographical characteristics for solar farm development. Forty polygons were created using the Thiessen (Voronoi) polygon method in ArcGIS indicating areas of influence around each point, representing closed landfill locations to understand the geographic extent of influence or service areas for each landfill site. The analysis revealed that the average elevation of 40 landfill sites ranges from 1.22 to 38.3 m, and the average slope ranges from 1.35 to 3.3 degrees. Additionally, the aspect, which indicates the direction of the slope, varied between 140.1 and 170.1 degrees. These factors are crucial for understanding surface drainage patterns and assessing the structural stability of the landfills for potential solar development (Szabó et al., 2017; Abdo et al., 2023). A mean slope of less than 5 degrees ensures that the land is nearly flat, reducing potential challenges related to solar panel installation and maintenance. The south-facing aspect maximizes sunlight exposure, crucial for

Table 14.1 Tolal solar energy potential of four Saskatchewan's census divisions

Division	*Area (km²)*	*Population density (km²)*	*Largest center*	*Eligible landfills number*	*Average solar radiation (Kwh/m²/day)*	*Total solar energy Potential (Kwh/day)*
9	14,851	2.4	Yorkton	10	4.51	722,280.16
10	11,922	1.4	Wynyard	5	4.59	343,098.62
6	17,363	15.9	Regina	4	4.65	319,146.71
11	16,602	19.6	Saskatoon	2	4.64	27,864.09

efficient solar energy generation. Additionally, a mean elevation of less than 150 m does not present significant barriers to construction or maintenance activities.

Only 21 identified areas are likely to be good candidates for future redevelopment based on the area calculated. Policymakers can use this simple check before considering using this land for solar farms. Where possible, a comprehensive analysis incorporating other influential factors such as environmental impact, soil stability, and proximity to urban infrastructure is recommended, since it is likely to yield a more accurate assessment of redevelopment potential, as noted by Nochian et al. (2016). Solar energy potential estimation of closed landfill sites involves consideration of key parameters, including the area of the landfills and the local solar radiation levels. Table 14.1 shows the total solar energy potential from the eligible landfills of each division. The solar energy potential, measured in kilowatt-hours per day (kWh/day), is calculated by combining the area of the landfill (in meter-square) with the average daily solar radiation (in kWh/m²/day).

The analysis indicates that Division 9, with 10 closed landfill sites, offers the highest solar energy potential among the four divisions considered. The Pelly landfill (Division 9), covering approximately 15.3 acres and receiving 4.38 kWh/m²/day of solar radiation, can generate an estimated 30,909 kWh/day. Larger sites, such as Lintlaw (Division 9), spanning 138 acres with 4.5 kWh/m²/day radiation, could produce around 272,085 kWh/day. Altogether, Division 9 has the potential to generate approximately 722,280 kWh/day from its landfills. In contrast, Division 11 has a potential for solar energy reuse for only two closed landfill sites. These estimates can guide the development of solar energy projects, optimizing the land previously used for landfilling to generate solar power.

The result suggests that these closed landfills present significant potential for solar farm development due to their favorable topography, large area, and electricity generation capability. Gungor and Dincer (2021) explored the potential of converting inactive landfills in Canada into solar farms, emphasizing the dual benefits of harnessing solar energy and capturing landfill gases to manage waste while generating renewable energy. Similarly, Niblick and Landis (2016) evaluated the suitability of closed landfills in the USA for utility-scale single-axis tracking photovoltaic systems and estimated that these sites could generate between 96.7 and 225 terawatt-hours of solar-based electricity annually. Leveraging closed landfill sites for solar energy production can contribute significantly to renewable energy goals and local energy security. The findings from this topographic analysis and solar energy potential can guide policy decisions on sustainable land use and renewable energy. Policymakers can prioritize repurposing suitable sites for solar projects, establish zoning regulations to encourage solar development, and consider economic and environmental impacts. Since environmental regulations and policies vary by province, project management must align with local legislative frameworks.

This underscores the need for a comprehensive, location-specific approach to developing solar farms on closed landfills, ensuring regulatory compliance and optimal energy generation outcomes. Additionally, for the most part, the significant existing engineering documentation for landfills, such as geological investigations, groundwater monitoring, and other environmental assessments offers an opportunity to refine decisions on optimal sites for solar farms. Furthermore, if a site has already undergone environmental assessment, there may be opportunities to leverage this preexisting site work, potentially streamlining the approval process for proposed solar projects. Finally, converting closed landfills into solar farms may indirectly support ongoing environmental compliance, as these sites would be reengineered to meet the requirements of the solar farm, ensuring they continue to meet postclosure environmental standards.

Limitations and future work

This study on closed landfills for solar farm development provides insights into the topographic analysis of these sites. However, topographic conditions alone do not fully capture the complex interplay of factors necessary for a comprehensive assessment. Key aspects such as soil stability, contamination levels, environmental impact, and infrastructure availability are not addressed in this study. Moreover, it does not include vector data analysis such as proximity to roads or electric transmission, assuming these facilities are already present. Furthermore, this GIS analysis does not consider socioeconomic factors such as living standards or climatic conditions, which can significantly influence the feasibility of solar farm projects. Additionally, reliance on online tools for calculating solar radiation provides valuable insights but may be limited by their generalized data inputs and assumptions.

For a more holistic evaluation, future research should incorporate a detailed soil analysis to ensure the stability and suitability of the ground for solar panel foundations. An environmental impact assessment is necessary to evaluate potential adverse effects on local ecosystems and propose mitigation measures. An economic feasibility study should assess the cost-effectiveness of converting closed landfills into solar farms, considering potential incentives and subsidies. Implementing pilot projects on selected sites would allow for the monitoring of performance and addressing any unforeseen challenges. Site-specific assessments, including detailed on-site measurements to complement satellite-based data, are essential for more accurate assessments of solar radiation. Advocating for supportive policies and incentives to promote solar energy adoption in the region is essential.

Conclusions

This chapter explores the potential of repurposing closed landfills in Saskatchewan as solar farms using GIS-based topographic analysis. Factors such as slope, aspect, and elevation were examined through DEM imagery to assess the technical suitability of these sites for solar energy projects. The analysis confirms that the mean values of these factors fall within the optimal range for solar energy harnessing, making closed landfills viable for sustainable development. Additionally, while 21 out of 40 landfills possess sufficient land area for redevelopment, the overall solar energy potential varies by division, with Division 9 showing the highest promise due to its greater number of closed landfills. By leveraging these sites for solar energy production, stakeholders can enhance renewable energy goals and support local energy needs, contributing to greenhouse gas reduction targets. Given that landfill sites are generally well investigated during their operating lifespan, policymakers and owners may be able to take advantage of existing environmental investigations to further understand where it is best to develop solar farms. The case study provides a foundation for further feasibility studies, including economic assessments, environmental impact analyses, and infrastructure planning. Addressing challenges such as cost, complexity, and institutional barriers is crucial for developing a strategic management system for repurposing closed landfills.

References

Abdo, H. G., Aljohani, T. H. D., Almohamad, H., Al-Dughairi, A. A., & Al-Mutiry, M. (2023). Sanitary municipal landfill site selection by integration of GIS and multi-criteria techniques for environmental sustainability in Safita area, Tartous governorate, Syria. *Environmental Science and Pollution Research*, 30(11), 30834–30854. https://doi.org/10.1007/s11356-022-24287-9.

Ali, S., Taweekun, J., Techato, K., Waewsak, J., & Gyawali, S. (2019). GIS based site suitability assessment for wind and solar farms in Songkhla, Thailand. *Renewable Energy*, 132, 1360–1372. https://doi.org/10.1016/j.renene.2018.09.035.

Arab, A. H., Taghezouit, B., Abdeladim, K., Semaoui, S., Razagui, A., Gherbi, A., Boulahchiche, A., & Mahammed, I. H. (2020). Maximum power output performance modeling of solar photovoltaic modules. *Energy Reports*, 6, 680–686. https://doi.org/10.1016/j.egyr.2019.09.049.

Assamoi, B., & Lawryshyn, Y. (2012). The environmental comparison of landfilling vs. incineration of MSW accounting for waste diversion. *Waste Management*, 32(5), 1019–1030. https://doi.org/10.1016/j.wasman.2011.10.023.

Buzra, U., & Serdari, E. (2023). A comparison analysis of different PV simulation tools using satellite data. *Electrical Engineering*, 105(4), 2485–2492. https://doi.org/10.1007/s00202-023-01814-6.

Chu, C. T., & Hawkes, A. D. (2020). A geographic information system-based global variable renewable potential assessment using spatially resolved simulation. *Energy*, 193, 116630. https://doi.org/10.1016/j.energy.2019.116630.

Department of Energy. (2024). *Solar radiation basics*. Retrieved August 1, 2024 from www.energy.gov/eere/solar/solar-radiation-basics.

Doorga, J. R., Rughooputh, S. D., & Boojhawon, R. (2019). Multi-criteria GIS-based modelling technique for identifying potential solar farm sites: A case study in Mauritius. *Renewable Energy*, 133, 1201–1219. https://doi.org/10.1016/j.renene.2018.08.105.

Environment and Climate Change Canada. (2022). *Reducing methane emissions from Canada's municipal solid waste landfills: Discussion paper*. Retrieved August 9, 2024 from www.canada.ca/en/environment-climate-change/services/canadian-environmental-protection-act-registry/reducing-methane-emissions-canada-municipal-solid-waste-landfills-discussion.html.

Environment and Climate Change Canada. (2023a). *Reducing Canada's landfill methane emissions: Proposed regulatory framework*. Retrieved August 9, 2024 from www.canada.ca/en/environment-climate-change/services/canadian-environmental-protection-act-registry/publications/reducing-landfill-methane-emissions.html.

Environment and Climate Change Canada. (2023b). *The Government of Canada invests in cleaning up Canada's landfill emissions*. Retrieved August 1, 2024 from www.canada.ca/en/environment-climate-change/news/2023/09/the-government-of-canada-invests-in-cleaning-up-canadas-landfill-emissions.html.

Environment and Climate Change Canada. (2024). *Solid waste diversion and disposal*. Retrieved August 9, 2024 from www.canada.ca/en/environment-climate-change/services/environmental-indicators/solid-waste-diversion-disposal.html.

Government of Canada. (2022). *Saskatchewan: Clean electricity snapshot*. Retrieved February 20, 2025 from www.canada.ca/en/services/environment/weather/climatechange/climate-plan/clean-electricity/overview-saskatchewan.html

Government of Saskatchewan. (2024). *Government of Saskatchewan*. Retrieved August 5, 2024 from https://geohub.saskatchewan.ca.

Government of Yukon. (2021). *Dawson City Solar Project connects to the Yukon electrical grid*. Retrieved August 1, 2024 from https://www.klondikedevelopment.com/dawson-city-solar-project-connects-to-the-yukon-electrical-grid.

Günen, M. A. (2021). A comprehensive framework based on GIS-AHP for the installation of solar PV farms in Kahramanmaraş, Turkey. *Renewable Energy*, 178, 212–225. https://doi.org/10.1016/j.renene.2021.06.078.

Gungor, B., & Dincer, I. (2021). Development of a sustainable community with an integrated renewable and waste to energy system for multiple useful outputs. *Journal of Cleaner Production*, 312, 127704. https://doi.org/10.1016/j.jclepro.2021.127704.

Hernandez, R. R., Easter, S. B., Murphy-Mariscal, M. L., Maestre, F. T., Tavassoli, M., Allen, E. B., … & Allen, M. F. (2014). Environmental impacts of utility-scale solar energy. *Renewable and Sustainable Energy Reviews*, 29, 766–779. https://doi.org/10.1016/j.rser.2013.08.041.

Huld, T. (2017). PVMAPS: Software tools and data for the estimation of solar radiation and photovoltaic module performance over large geographical areas. *Solar Energy*, 142, 171–181. https://doi.org/10.1016/j.solener.2016.12.014.

Jacob, B. A., & Ayers, M. R. (2018). Landfill solar: Trash to treasure. *Energy Engineering*, 115(5), 37–47. https://doi.org/10.1080/01998595.2018.12027707.

Kabir, E., Kumar, P., Kumar, S., Adelodun, A. A., & Kim, K. H. (2018). Solar energy: Potential and future prospects. *Renewable and Sustainable Energy Reviews*, 82, 894–900. https://doi.org/10.1016/j.rser.2017.09.094.

Laner, D., Crest, M., Scharff, H., Morris, J. W., & Barlaz, M. A. (2012). A review of approaches for the long-term management of municipal solid waste landfills. *Waste Management*, 32(3), 498–512. https://doi.org/10.1016/j.wasman.2011.11.010.

Melhem, R., & Shaker, Y. (2023). Optimum tilt angle and solar radiation of photovoltaic modules for gulf collaboration council countries. *International Journal of Energy Research*, 2023(1), 8381696. https://doi.org/10.1155/2023/8381696.

Moodliar, L., & Davidson, I. E. (2023). Upcycling trash into cash through repurposing the Bisasar Road Landfill Site into a solar PV and energy storage site in eThekwini municipality. In *Future energy: Challenge, opportunity, and sustainability* (pp. 25–36). Springer. https://doi.org/10.1007/978-3-031-33906-6_3.

Munsell, D. R. (2013). *Closed landfills to solar energy power plants: Estimating the solar potential of closed landfills in california*. Order No. 1538728. Retrieved from https://login.libproxy.uregina.ca:8443/login?url=https://www.proquest.com/dissertations-theses/closed-landfills-solar-energy-power-plants/docview/1400483911/se-2

Nai, C., Tang, M., Liu, Y., Xu, Y., Dong, L., Liu, J., & Huang, Q. (2021). Potentially contamination and health risk to shallow groundwater caused by closed industrial solid waste landfills: Site reclamation evaluation strategies. *Journal of Cleaner Production*, 286, 125402. https://doi.org/10.1016/j.jclepro.2020.125402.

National Renewable Energy Laboratory. (2024). *PVWatts calculator*. Retrieved August 1, 2024 from https://pvwatts.nrel.gov.

Natural Resources Canada. (2023). *Solar Power Saskatchewan*. Retrieved August 1, 2024 from www.energyhub.org/saskatchewan.

Niblick, B., & Landis, A. E. (2016). Assessing renewable energy potential on United States marginal and contaminated sites. *Renewable and Sustainable Energy Reviews*, 60, 489–497. https://doi.org/10.1016/j.rser.2015.12.045.

Nochian, A., Tahir, O. M., Maulan, S., & Mikaeili, A. R. (2016). A review of systematic approach for sustainable redevelopment of a closed landfill site. *Jurnal Teknologi*, 78(5). https://doi.org/10.11113/jt.v78.8318.

Okunlola, A., Davis, M., & Kumar, A. (2022). The development of an assessment framework to determine the technical hydrogen production potential from wind and solar energy. *Renewable and Sustainable Energy Reviews*, 166, 112610. https://doi.org/10.1016/j.rser.2022.112610.

Pesâkâstêw Solar Project. (2024). *Pesâkâstêw Solar Project*. Retrieved August 9, 2024 from www.pesakastewsolarproject.ca.

Quesada, G., Guillon, L., Rousse, D. R., Mehrtash, M., Dutil, Y., & Paradis, P. L. (2015). Tracking strategy for photovoltaic solar systems in high latitudes. *Energy Conversion and Management*, 103, 147–156. https://doi.org/10.1016/j.enconman.2015.06.041.

Rediske, J., et al. (2020). *Site suitability analysis for solar farms using the geographic information system*. Retrieved from https://link.springer.com/article/10.1007/s10098-020-02018-3.

Richter, A., Ng, K. T. W., & Karimi, N. (2019). A data driven technique applying GIS, and remote sensing to rank locations for waste disposal site expansion. *Resources, Conservation and Recycling*, 149, 352–362. https://doi.org/10.1016/j.resconrec.2019.06.013.

Singh, V., Rayal, I., Sharma, H., Dwivedi, C., & Singh, B. (2020). Solar radiation and light materials interaction. In *Energy saving coating materials* (pp. 1–32). Elsevier. https://doi.org/10.1016/B978-0-12-822103-7.00001-7.

Sliusar, N., Filkin, T., Huber-Humer, M., & Ritzkowski, M. (2022). Drone technology in municipal solid waste management and landfilling: A comprehensive review. *Waste Management*, 139, 1–16. https://doi.org/10.1016/j.wasman.2021.12.006

Solar FlexRack. (2022). *Solar FlexRack reaches 1,000 solar installations worldwide*. PR Newswire. Retrieved August 9, 2024 from www.prnewswire.com/news-releases/solar-flexrack-reaches-1-000-solar-installations-worldwide-301481130.html.

Srivastava, V., Ismail, S. A., Singh, P., & Singh, R. P. (2015). Urban solid waste management in the developing world with emphasis on India: Challenges and opportunities. *Reviews in Environmental Science and Bio/Technology*, 14, 317–337. https://doi.org/10.1007/s11157-014-9352-4.

Stohr, C., & Filippini, H. (2018). Enhanced field inspections of closed landfills using aerial orthophotography in Illinois. *Journal of Hazardous, Toxic, and Radioactive Waste*, 22(1), 04017022. https://doi.org/10.1061/(ASCE)HZ.2153-5515.0000372

Sukumaran, S., Sudhakar, K., Yusop, A. F., Kirpichnikova, I., & Cuce, E. (2022). Solar farm: Siting, design and land footprint analysis. *International Journal of Low-Carbon Technologies*, 17, 1478–1491. https://doi.org/10.1093/ijlct/ctac107.

Szabó, S., Bódis, K., Kougias, I., Moner-Girona, M., Jäger-Waldau, A., Barton, G., & Szabó, L. (2017). A methodology for maximizing the benefits of solar landfills on closed sites. *Renewable and Sustainable Energy Reviews*, 76, 1291–1300. https://doi.org/10.1016/j.rser.2017.03.117.

Tansel, B., Varala, P. K., & Londono, V. (2013). Solar energy harvesting at closed landfills: Energy yield and wind loads on solar panels on top and side slopes. *Sustainable Cities and Society*, 8, 42–47. https://doi.org/10.1016/j.scs.2013.01.004.

US Environmental Protection Agency. (2023). Best practices for siting solar photovoltaics on municipal solid waste landfills. Retrieved August 1, 2024 from www.epa.gov.

US Geological Survey. (2024). *EarthExplorer*. Retrieved August 15, 2024 from https://earthexplorer.usgs.gov.

Weng, Y. C., Fujiwara, T., Houng, H. J., Sun, C. H., Li, W. Y., & Kuo, Y. W. (2015). Management of landfill reclamation with regard to biodiversity preservation, global warming mitigation and landfill mining: experiences from the Asia–Pacific region. *Journal of Cleaner Production*, 104, 364–373. https://doi.org/10.1016/j.jclepro.2015.05.014.

Williams, H. J., Peng, H., Goosay, O., & Zhang, K. M. (2023, June). Brownfields to brightfields: The potential for landfill solar redevelopment in New York State. In *2023 IEEE 50th photovoltaic specialists conference* (pp. 1–3). IEEE. https://doi.org/10.1109/PVSC48320.2023.10359589.

Zahari, M. S. M., Jaafar, I., Ismail, S., Abu Amr, S. S., Aziz, H. A., & Hung, Y. T. (2022). Landfill after-care management plan. In *Solid waste engineering and management* (vol. 2, pp. 659–706). Springer. https://doi.org/10.1007/978-3-030-89336-1_10.

Zuberi, M. J. S., & Ali, S. F. (2015). Greenhouse effect reduction by recovering energy from waste landfills in Pakistan. *Renewable and Sustainable Energy Reviews*, 44, 117–131. https://doi.org/10.1016/j.rser.2014.12.028.

Part 5

Emerging waste management

15 Emerging solid wastes and arising environmental impacts

Delezia S. Singh

Introduction

Solid waste management represents a crucial and ongoing environmental issue, as countries like Canada continue to deal with evolving patterns of resource consumption and increasing waste production. New types of solid wastes are rapidly developing alongside changing lifestyles and rapid obsolescence of "older" technology, so it is no surprise that these represent prominent sources of emerging contaminants of concern today. Solid waste generation in Canada reflected a notable increase of 5.3 million tons from 2002 to 2020, mounting to a total of 36.0 million tons (ECCC, 2023). Provinces like Manitoba with growing populations reportedly generate over 900,000 tons of municipal solid waste yearly and densely populated Ontario, Quebec, Alberta and British Colombia produce over 2 million tons in overall wastes biyearly, inclusive of residential and nonresidential solid wastes for disposal (Statistics Canada, 2023). Management challenges associated with such high waste generation across Canada are evident for recent years, like 2020, wherein 72.5% of solid wastes were sent for disposal and only 27.5% diverted (ECCC, 2023).

Despite advances in technological improvements, landfilling is still one of the main approaches for solid waste disposal within Canada (Nazari et al., 2021). Waste disposal via this method or incineration reflected a 2.0 million ton increase in 2020 across Canada (ECCC, 2023). Legacy chemicals like persistent organic pollutants, fluorinated compounds and brominates (BTEX/benzene, toluene, ethyl benzene, and xylenes, per- and polyfluoroalkyl substances, pesticides, polychlorinated biphenyls, phenols, sulfamic acid and perchlorate compounds) that are highly toxic to human health and aquatic life, have been detected in environmental samples associated with solid waste landfills in North America and across provinces in Canada (Ochs et al., 2024). Coupled with this are distinct challenges to solid waste management for Canadian northern communities, stemming from permafrost conditions, deficiency in appropriate cover material for landfill sites and high logistical costs associated with transportation of recyclable materials (Muir & Galarneau, 2021). This chapter will therefore highlight emerging solid wastes that have progressively contributed to the overall waste stream, along with the problematic contaminants they produce and the resulting environmental impacts.

DOI: 10.4324/9781003534129-21

Emerging solid wastes: Progressive contributors to the overall waste stream

Rapid technological development, high demands and consumption of electricals and electronic equipment have enhanced the volume of end-of-life electrical/electronic (e-) waste and unfair exportation of e-wastes (Erdiaw-Kwasie et al., 2024). It is anticipated that the generation of e-waste, especially from energy storage devices, will continue to significantly increase. This is attributed to these devices being integral components of most electronic products and Internet of Things (IoT)-based systems. Simultaneously, the enhancement of electronic trash containing hazardous components have created multifaceted environmental and human health problems, along with placing pressure on current waste diversion and management systems (Sahle-Demessie et al., 2021). In response, material recovery from mining electronic-based wastes is gaining recognition due to its economic value and provision of commercial and income opportunities (Sahle-Demessie et al., 2021; Erdiaw-Kwasie et al., 2024).

However, use of informal salvaging methods like incineration, manual disassembly and chemical processing procedures involving acidic leaching can still generate toxic products, including heavy metals, furans, volatile organic, aromatic and brominated compounds, for example, lead, benzene, benzofuran and bromophenol produced from pyrolysis thermal treatment of mixed e-wastes (Sahle-Demessie et al., 2021). Recyclable electronics can therefore be quite challenging, due to their problematic components with diverse physicochemical properties being present in e-waste streams, including toxic metals (mercury, phosphor coatings of cathode ray tubes, lead content from batteries, lead silicate glass from scrapped cathode ray tubes), capacitor-derived polychlorinated biphenyls, brominated flame retardants and plastics (Ilankoon et al., 2018).

The generation of these contaminants from e-wastes limits traditional recycling processes of reusable materials, especially those with plastic-based components. Intrinsically, plastic waste pollution continues to be a persistent, almost irresolvable issue, although recent research interests have revived around their micropollutant relatives that have become environmentally prevalent, that is, microplastics. These plastic fragments can originate from primary manufacturing processes but can be secondarily derived from a variety of plastic product wastes (Haque & Fan, 2022). Due to the ubiquitous presence of macroplastic (over 25 mm), mesoplastic (25–5 mm), microplastic (less than 5 mm particle size) and nanoplastic (less than 1 μm) contaminants worldwide, plastics represent one of the most challenging groups of solid wastes for management to address in the 21st century. One such "petite" plastic problem takes the form of glitter, which collectively comprise of tiny, shiny and decorative particles, widely produced and used for aesthetic and ornamental purposes in entertainment, cosmetic and fashion industries (makeup, bags, clothing, carnival costumes etc.). Significant concerns surrounding glitter-imbued wastes have recently emerged, due to the polyethylene terephthalate/aluminum constituents of synthetic glitter, their environmental persistence and the wide range of common commercial products that contain them (Yurtsever, 2019).

Despite the restrictive measures placed on plastic glitter usage by regulatory bodies like the European Commission, and the innovation of biodegradable forms (bioglitter) containing environmentally sensitive plant-based/cellulose or mica cores, there is still the matter of long-term environmental accumulation of glitter microplastics and the unrecoverable/unrecyclable nature of used plastic glitters (Yurtsever, 2019). Also, much is still unknown about the environmental impacts of the biotypes—current ecotoxicological research has already demonstrated acute toxicity of growth (biomass and chlorophyll) impairments in *Lemna minor* from short-term exposures of modified cellulose bioglitter (Boots et al., 2023).

The Covid-19 outbreak has played a significant role in exacerbating plastic burdens on habitats and biota, since the ongoing pandemic spurred the enhanced usage and disposal of plastic products, driven by safety concerns, increased online shopping, takeout, food home delivery and panic shopping. Since the outbreak, 1.6 million tons of plastic waste/day were generated globally (Benson et al., 2021). Polyethylene terephthalate is a notable contributor and is prevalent in single-use food packaging items like clamshell containers, salad domes, snack trays and bottle caps (Haque & Fan, 2022). Global panic also heightened demands and spiked production for personal protective equipment (PPE), like single-use face masks and gloves. Accordingly, unplanned waste management coupled with the unusually high generation of pandemic-related plastic waste would have led to unintentional microplastic release from these sources.

The pandemic also indirectly contributed to the solid waste streams of the construction, renovation and demolition (CR&D) sector, due to a rise in home improvement and renovation activities during the lockdown periods (Peula et al., 2023). Pre-Covid-19, growing populations, enhanced industrialization and urban expansion were already fueling a rapid increase of CR&D wastes. Discarding of CR&D wastes is still mainly done through landfill deposition, resulting in pollution from hazardous leachates, hydrogen sulfide emissions from anaerobic decomposition of debris, trace metals, microplastics from foam insulation and other polymer applications, soluble sulfates, acids, residual polycyclic aromatic hydrocarbons (PAHs) and gypsum dust particles (Cook et al., 2022; Laadila et al., 2022). Furthermore, the valorization of the CR&D solid waste division is difficult to valorize due to the presence of these toxic compounds that could be otherwise released during diversion processes. Apart from direct activities, severe weather events driven by climate change have also indirectly produced disaster-related CR&D waste from infrastructural and building damages. Cyclones, hurricanes and other extreme disasters (e.g., earthquakes) can contribute highly diverse and atypical varieties of solid wastes, abruptly, and at abnormally high volumes (Chandrappa & Das, 2024). Flooding, tsunamis and urban stormwater runoff may extensively spread solid debris and garbage via flowing water, clog drainage systems and disrupt stream morphology, while demolition debris from damaged or collapsed buildings can impede emergency services, and wildfires can produce ash, particulate matter and charred solid wastes from wood, mixed metals, burnt cars and buildings (Chandrappa & Das, 2024).

Another emerging group of contaminants, pharmaceuticals and personal care products (PPCPs) have been on the research radar in recent decades. PPCPs encompass over-the-counter, unused or expired pharmaceutical drugs and natural health care products. Globally, their high production, consumption rates and continuous environmental discharge make PPCP contaminants considerably concerning. With the onset of Covid-19, scientific interests surrounding these xenobiotics have heightened. The pandemic would have influenced the extraordinary increase in personal care product consumption of soaps, sanitizers, detergents and other cleaning products, and indirectly reduced recycling collection (Mohamed et al., 2022; Marteinson et al., 2023), likely resulting from high volumes of wastes generated from these items, for example, packaging, plastic-based containers. Additionally, a surge in the usage of nonprescription drugs would have been attributed to over-the-counter pharmaceuticals, like ibuprofen and paracetamol (acetaminophen), which were being recommended by public health authorities as effective medications for managing less severe symptoms of the virus (Karami et al., 2023). Heightened pharmaceutical consumption was also driven by drug misuse in the form of prophylactic self-medicating performed before vaccination or "preventative" panic dosing to avoid contracting Covid-19 (Yáñez et al., 2021).

Environmental contamination from the solid-state drug forms (e.g., tablets, capsules) would have been likely worsened by the improper disposal of expired or unused medications, resulting from drug stockpiling by consumers during the pandemic. Altogether, this is aggravated by the marked increase in new veterinary and human-based pharmaceutical products entering the Canadian drug marketplace, advances in drug research and development, increased accessibility to healthcare and high medical demands from growing populations. Pharmaceutical disposal programs are useful initiatives for reducing improper disposal of healthcare products, expired and unwanted drugs (Gagnon, 2009). These operate with the help of community pharmacies and collection depots throughout provinces and are regulatory in British Columbia and New Brunswick (Gagnon, 2009). Though these are longstanding proactive efforts, only a fraction of all pharmaceutical wastes could be captured, considering 30% of Canadian households reportedly still used uncontrolled disposal methods for expired or unused medications, for example, regular garbage disposal, flushing down the sewer, drains or sinks (Statistics Canada, 2011).

Contaminants of concern associated with emerging solid wastes

Since solid wastes comprise such a miscellaneous array of evolving materials and debris, these continue to be central sources of contaminants, several of which have only recently appeared on the research radar, having garnered attention within later decades. Emerging contaminants or contaminants of emerging concern (CECs) can be naturally occurring or synthetic-based chemicals (recently synthesized or long-term existing substances), even biological entities such as microorganisms, and having the potential to be present in the environment but not commonly monitored (Wang et al., 2024). There are numerous types of these toxic substances derived from heterogenous solid wastes and their processing/treatment methods (Table 15.1),

Table 15.1 Contaminants of concern generated from solid waste disposal methods, according to environmental matrices in which they are released or contained

Disposal method	*Contaminants within different matrices*			
	Atmospheric	*Aquatic*	*Terrestrial/soil*	*Biological*
Landfilling and open dumpsites[a]	GHGs; H_2S; gypsum/sulfur dust particles (CR&D); PAC; fugitive VOCs; CFCs	Leachate; M&NPs (glitter, Covid-19 masks, gloves and PPE, food packaging, CR&D foamed insulation, building polymer debris); natural disaster debris; HMs (e-waste and CR&D); PPCPs & byproducts; legacy compounds (organobromines, PFAS, PCBs, pesticides); sulfates (CR&D); biohazardous microorganisms[b]	Leachate; HMs (e-waste); M&NPs; pesticides; biohazardous microorganisms[b]; HGT	M&NP in animal (aquatic and terrestrial), plant (bioadhesion) and human compartments (blood, breast milk and maternal/fetal transfer); bioaccumulated PFAS, PCBs, HMs; infectious microorganisms; HGT
Incineration	Fugitive GHGs; VOCs; acids; CO; PCDD/Fs, PACs; halocarbons; metallic particles	Atmospheric fluxes (wet deposition) of organic contaminants, PM, soot, inorganic ash with metallic and mineral particles	Atmospheric fluxes (dry deposition) of organic contaminants, PM, soot, inorganic ash with metallic and mineral particles	Noxious gases (H_2S, PCDD/Fs, volatile PFAS); soot; dust; inorganic PM; fly ash with metallic and mineral particles (inhalation)
Recycling (waste diversion)	GHGs (negligible)	M&NPs; PFAS; HMs; phenolic compounds (liquid waste streams)	M&NPs; PCDD/Fs; PFAS; organobromines (agricultural applications)	M&NPs; PCDD/Fs; PFAS; organobromines (dietary uptake)

(*Continued*)

Table 15.1 (Continued)

Disposal method	*Contaminants within different matrices*			
Composting and WtE (diversion)	Bioaerosols; fugitive GHGs; hazardous flue gas (improper handling); dioxins	Biohazardous microorganisms[b]; leachates from WtE ash and slag (HMs, PFAS, PCDD/Fs); M&NPs (compostable items, bioplastics)	Biohazardous microorganisms[b]; HGT; WtE ash and slag (HMs, PFAS, PCDD/Fs); M&NPs (compostables, bioplastics)	Biohazardous microorganisms[b] including drug-resistant bacteria; occupational/ residential flue gas exposures (WtE)

Source: This table was adapted and modified from Giusti (2009) to include emerging and more recent contaminants.

Notes

[a] Includes improper landfill treatment processes and unregulated dumping.

[b] Biohazardous microorganisms = viruses, antibiotic/drug-resistant bacteria, parasites and fungi.

GHGs = greenhouse gases (direct and indirect) referring to methane (CH_4), carbon dioxide (CO_2), sulfur dioxide (SO_2) and nitrogen oxides (NO_x); PACs = polycyclic aromatic compounds; CFCs = chlorofluorocarbons; HMs = heavy metals; PM = particulate matter; M&NPs = micro- and nanoplastics; VOCs = volatile organic compounds; PFAS = per- and polyfluoroalkyl substances; PCBs = polychlorinated biphenyls; CO = carbon monoxide; PCDD/Fs = polychlorinated dibenzo-p-dioxins and -furans; HGT = horizontal genetic transfer; WtE = Waste-to-Energy.

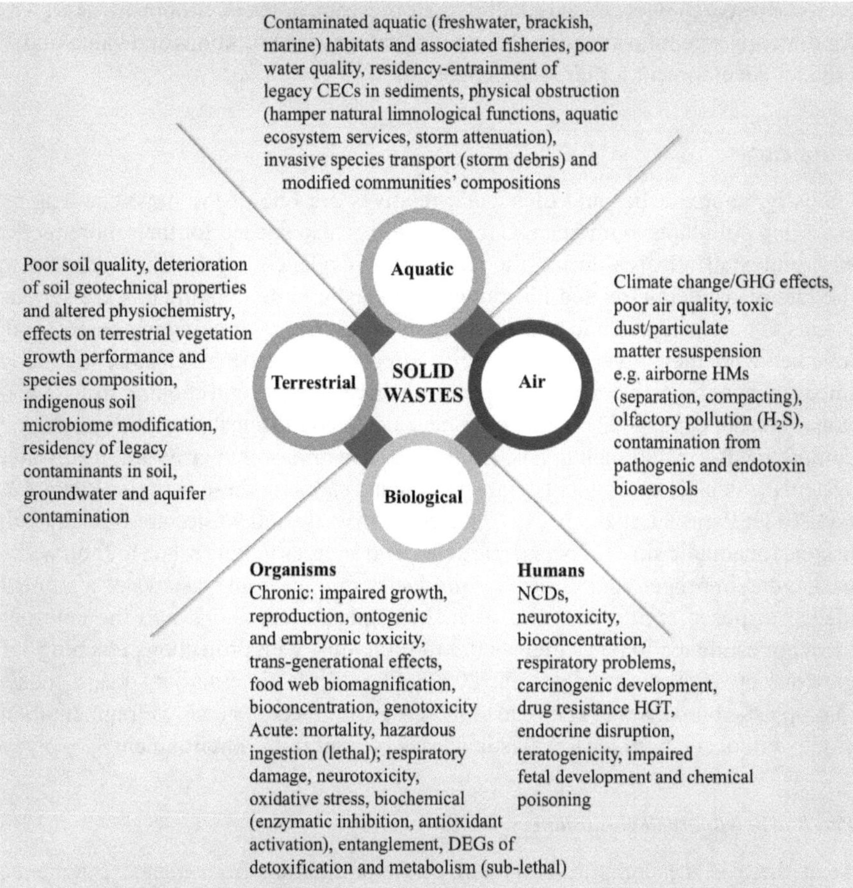

Figure 15.1 Environmental impacts associated with different matrices resulting from CECs generated from solid wastes.

Note: NCDs refer to noncommunicable diseases and DEGs are differentially expressed genes.

Sources: Impacts listed within this figure were adapted from those discussed by Chandrappa and Das (2024), to include effects from contaminants that have grown in recent concern. This figure and its concept design was originally created by the author of this chapter.

eliciting both acute and long-term toxicity effects across many levels of biological organization, including genotoxicity, carcinogenicity and impairment of ecosystem services (Figure 15.1). Several key CEC groups have gained increasing awareness within the government, scientific and public domains: micro- and nanoplastics, PPCPs, microorganisms, greenhouse gases (GHGs) and air pollutants, heavy metals and persistent organic pollutants, like per- and polyfluoroalkyl substances (PFAS). The relatively more recent contaminant groups and their associated solid waste-related sources will be discussed. These CECs warrant highlight in order to increase research interests in reducing their presence in the environment, as well as for directing scientific attention toward innovative waste solutions and value-added product development for circular economies.

Microplastics

Globally, microplastics and their nano-relatives are one of the most challenging emerging pollutants to mitigate. Consideration is also needed for their more recent environmental releases during the Covid-19 prevalence, markedly in relation to the fate of pandemic-related plastic wastes. Plastic particle pollutants are serious threats to human health due to their pervasive presence in drinking water and food, together with their diversity of toxicity effects on aquatic organisms, including impairment of digestive functioning, reproductive capacity and choking from ingestion (Parashar et al., 2023). These have been detected in almost all biotic and abiotic compartments, including humans, Arctic ice, terrestrial soil, marine and freshwater systems which act as recipients, transporters and sinks for these particles (Haque & Fan, 2022; Parashar et al., 2023). Terrestrial-derived solid wastes are fundamental origins for aquatic sinks of plastic particles, and sources largely relate to food waste packaging, improper landfilling, end-products (sludge) from wastewater treatment plants, disposed textiles and tire wear. Microplastic emissions into the environment are estimated to be as high as 0.8 million tons, with projections reaching 3.1 gigatons by 2050 (Schwarz et al., 2023). Should plastic emissions cease today, microplastic burdens will continue to increase within ecosystems, as fragmentation would still occur from legacy plastic wastes present in the environment.

Per- and polyfluoroalkyl substances

Then there is the complex nature of co-contamination from plastic fragments, whereby these pose combined toxicity risks from being originally incorporated with a cocktail of contaminants (e.g., PFAS, heavy metals) during the manufacturing process of commercial products. Following the end-life or unregulated disposal of plastic-based items, the resulting microplastics can serve as "trojan horses" for other harmful contaminants because of the high sorption capacity of their surface areas for organic and inorganic micropollutants. For instance, plastic additives like PFAS and other legacy contaminants were detected in Canadian Arctic seabirds, the northern fulmar and black-legged kittiwake (Sühring et al., 2022), with PFAS being recognized as contaminants of emerging Arctic concern. PFAS-adsorbed microplastics can alter the former group's bioavailability, increase their degradative resistance and intensify toxicity in aquatic biota from combined exposure (Parashar et al., 2023). PFAS represent a fluorinated class of more than 4,700 synthetic chemicals known for their extensive use in various commercial and industrial applications (Schwartz-Narbonne et al., 2023). These bioactive chemicals include surfactants in stain repellent products (such as carpets, rainwear, textiles and rubber), oil- and grease-resistant/nonstick food packaging and surface protectors (Glüge et al., 2020). Food packaging samples (including sandwich, dessert and bread wrappers, popcorn serving bags and molded compostable fiber bowls) collected from Toronto-based food retailers revealed 22 detected PFAS, with the highest levels of fluorine detected from compostable bowls (Schwartz-Narbonne et al., 2023); further emphasizing that even solid wastes from "pro-environmental" single-plastic substitutions are not shy of producing dangerous contaminants.

Legacy PFAS are of significance due to their strong carbon–fluorine bonds, which provide high chemical stability and resistance to oxidative and degradative processes (Parashar et al., 2023). Such notorious chemical members of the long-chain PFAS isomers (PFHxS, PFCAs and PFOS) have been estimated to have long half-lives in humans (Zhang et al., 2023) and bioaccumulative effects across diverse trophic levels, from phytoplankton to fish, Arctic seals, porpoises and polar bears. Despite restrictions and voluntary phaseouts of problematic PFAS, precursor PFAS families are still in regulatory use—such as in coatings for medical PPE, pharmaceutical equipment, semiconductor and electrical components—and can still degrade into more persistent species (Glüge et al., 2020).

Pharmaceuticals and personal care products

Alternatively, pseudo-persistent PPCPs have been suggested to have potentially greater environmental persistence, compared to organic pollutants, because of their almost constant source releases (even considering their transformation during photodegradation, biodegradation and sorption processes). Leachate from the treatment of municipal solid wastes represents a significant source of PPCPs entering the environment (Zhang et al., 2023). When exposed to various environmental conditions, PPCPs and associated active ingredients contained within waste streams and leachates will break down into equally or more toxic metabolites and byproducts. PPCPs consequently pose long-term risks to nontarget aquatic species, soil organisms and can potentially cause various human health problems, including genotoxic and carcinogenic (cancer) effects. This is especially notable for medical drugs, since the evolutionary conservation of target molecular proteins and mechanisms across species, combined with the specific action modes of pharmaceutical compounds, heighten the risk of pharmacological activation and high-affinity interactions in nontarget biota (Fabbri & Franzellittiyz, 2016). As an example, NSAID drugs, like ibuprofen, diclofenac and ketoprofen have been associated with cardiac impairments and embryonic teratogenicity in zebra fish, and antibiotics have reportedly induced toxic effects in aquatic biota, such as green algae, cyanobacteria, crustaceans, duckweed and fathead minnow species (Schwartz et al., 2021).

Air pollutants

Aside from aquatic impacts, decomposing municipal solid waste can produce important gaseous pollutants, including GHGs, like methane and carbon dioxide, which have strong global warming potentials. In addition, solid waste emissions can contribute other harmful ozone-depleting halogenic and volatile organic compounds (VOCs) which also cause negative climate impacts. Landfill gas reportedly contained 63 non-methane VOC species, along with trace gas concentrations at 1–10,000 μg/L (Manheim et al., 2021). Some of the main VOCs are benzene, toluene, ethylbenzene, and xylenes (also known as BTEX) which collectively, and individually, represent human carcinogens at low concentrations. Lakhouit and Alsulami (2020) investigated BTEX concentrations from emitted biogas collected from the Saint Nicephore landfill, Quebec, and found that the detected compounds (except benzene) were within unacceptable levels that placed excess risk of cancer development on landfill workers and residents living nearby.

Notwithstanding modern engineered mitigations, there is still the complication of fugitive landfill gases and the release of air pollutants from full-scale solid waste incineration; the latter being another main approach for dealing with solid waste generation. In Canada, various incinerator types are employed, including waste-to-energy facilities, hazardous waste and biomedical incinerators. Countrywide standards and other best management practices are developed by the Canadian Council of Ministers of the Environment to mitigate dioxins, furans and the heavy metal, mercury, emitted from waste incinerations. However, each removal technology is variable in performance for ignition end-products, for example, dust and acidic gas wet scrubbers can unintentionally magnify dibenzo-p-dioxins and -furan concentrations contained in flue gas output (Lin et al., 2020). Unintentional incinerations can also be accountable, such as the case of an ongoing fire at the municipal solid waste landfill of Iqaluit, Nunavut that generated elevated polycyclic aromatic compounds from the diverse range of housed solid wastes (Weichenthal et al., 2015).

Biological contaminants

In addition to synthetic pollutants, biological contaminants associated with solid wastes can also be worrying. Though indigenous microorganisms have shown promising utility in waste management for metabolizing harmful compounds (e.g., heavy metals) through bioaccumulation and intracellular sequestration, improper disposal, treatment or accidental spills of solid wastes from medical and agricultural sectors can result in biohazard contamination (fungi, algae, viruses, protozoa and bacteria) of the ambient surroundings. Another problem lies in the evolution of pathogenic and antibiotic-resistant assemblages noted for undesirable microbes. Pathogenic (e.g., *Escherichia coli*, *Salmonella* sp.) and antibiotic-resistant bacteria can be present in landfill leachate, and their proliferation is promoted by other contaminants (pharmaceuticals) present in solid wastes that provide favorable environments and selection pressures. There is particular focus on antibiotic resistance genes and genetic elements (plasmids, integrons and transposons) that can undergo horizontal genetic transfer (HGT) outward to other microbial strains, resulting in biocontamination problems and dissemination of drug resistance across the natural and clinical environments (Tripathi & Tripathi, 2017).

Conclusion

Key actions in reducing environmental impacts from emerging solid wastes reside in reducing the volumes produced at the manufacturing and consumer levels. Primarily, the use of virgin natural resources needs to be reduced during production, alongside decreasing waste generation through redirecting and reusing recyclable wastes from landfills (Nazari et al., 2021). This is cost-effective for dealing with landfill hazardous wastes while closing the loop of the product lifecycle. Product stewardship programs and a Canada-wide action plan for extended producer responsibility strategy have been utilized for managing the end-of-life phase of products (Nazari et al., 2021). The former approach delegates responsibilities to municipal governments, provinces or territories while the latter favors greater producer responsibility toward the life cycle of the products, incentivizes waste reduction at the source and promotes proenvironmental product design (ECCC, 2017). Similarly, integrated solid waste management programs can tackle prevention, recycling/treatment and disposal, coupled with life cycle assessments that evaluate environmental impacts associated with products' life cycles, processes and services (Chandrappa & Das, 2024). As Canada strives to meet its waste reduction targets and circular economy goals, the increasing complexity and environmental issues surrounding emerging solid wastes necessitates a paradigm shift toward these approaches for promoting sustainable solid waste management.

References

Benson, N. U., Bassey, D. E., & Palanisami, T. (2021). Covid pollution: Impact of Covid-19 pandemic on global plastic waste footprint. *Heliyon*, 7(2), article e06343. https://doi.org/10.1016/J.HELIYON.2021.E06343.

Boots, B., Green, D. S., Olah-Kovacs, B., De Falco, F., & Lupo, E. (2023). Physical and chemical effects of conventional microplastic glitter versus alternative glitter particles on a freshwater plant (Lemnaceae: *Lemna minor*). *Ecotoxicology and Environmental Safety*, 263, article 115291. https://doi.org/10.1016/j.ecoenv.2023.115291.

Chandrappa, R., & Das, D. B. (2024). *Solid waste management: Principles and practice* (2nd ed.). Springer Nature. https://doi.org/10.1007/978-3-642-28681-0.

Cook, E., Velis, C. A., & Black, L. (2022). Construction and demolition waste management: A systematic scoping review of risks to occupational and public health. *Frontiers in Sustainability*, 3, article 924926. https://doi.org/10.3389/frsus.2022.924926.

Environment and Climate Change Canada. (2017). *Overview of extended producer responsibility in Canada*. Environment and Climate Change Canada, Government of Canada. www.canada.ca/en/environment-climate-change/services/managing-reducing-waste/overview-extended-producer-responsibility.html.

Environment and Climate Change Canada. (2023). *Canadian environmental sustainability indicators: Solid waste diversion and disposal*. Environment and Climate Change Canada, Government of Canada. www.canada.ca/content/dam/eccc/documents/pdf/cesindicators/solid-waste/2024/solid-waste-diversion-disposal-en.pdf

Erdiaw-Kwasie, M. O., Abunyewah, M., & Baah, C. (2024). A systematic review of the factors: Barriers, drivers, and technologies—Affecting e-waste urban mining: On the circular economy future of developing countries. *Journal of Cleaner Production*, 436, article 140645. https://doi.org/10.1016/j.jclepro.2024.140645.

Fabbri, E., & Franzellittiyz, S. (2016). Human pharmaceuticals in the marine environment: focus on exposure and biological effects in animal species. *Environmental Toxicology and Chemistry*, 35(4), 799–812. https://doi.org/10.1002/etc.3131.

Gagnon, E. (2009). *Pharmaceutical disposal programs for the public: A Canadian perspective*. Health Canada. https://citeseerx.ist.psu.edu/document?repid=rep1&type=pdf&doi=5a9536eb8067ffeedbcbae570ef6c79e2f9bc1df

Giusti, L. (2009). A review of waste management practices and their impact on human health. *Waste Management*, 29(8), 2227–2239. https://doi.org/10.1016/J.WASMAN.2009.03.028.

Glüge, J., Scheringer, M., Cousins, I. T., Dewitt, J. C., Goldenman, G., Herzke, D., Lohmann, R., Ng, C. A., Trier, X., & Wang, Z. (2020). An overview of the uses of per- and polyfluoroalkyl substances (PFAS). *Environmental Science: Processes and Impacts*, 22(12), 2345–2373. https://doi.org/10.1039/d0em00291g.

Haque, F., & Fan, C. (2022). Prospect of microplastic pollution control under the "new normal" concept beyond Covid-19 pandemic. *Journal of Cleaner Production*, 367, article 133027. https://doi.org/10.1016/j.jclepro.2022.133027.

Ilankoon, I. M. S. K., Ghorbani, Y., Chong, M. N., Herath, G., Moyo, T., & Petersen, J. (2018). E-waste in the international context: A review of trade flows, regulations, hazards, waste management strategies and technologies for value recovery. *Waste Management*, 82, 258–275. https://doi.org/10.1016/J.WASMAN.2018.10.018.

Karami, S., Asonye, C., Pinnow, E., Pratt, V., McCulley, L., Dwumfour, N., & Zhou, E. H. (2023). Trends in pediatric nonprescription analgesic/antipyretic exposures during the Covid-19 pandemic. *Clinical Toxicology*, 61(3), 190–199. https://doi.org/10.1080/15563650.2022.2158847.

Laadila, M. A., LeBihan, Y., Caron, R. F., & Vaneeckhaute, C. (2022). Physical and chemical characterization of construction, renovation and demolition waste in the Quebec province. *Cleaner Waste Systems*, 1, article 100002. https://doi.org/10.1016/J.CLWAS.2022.100002.

Lakhouit, A., & Alsulami, B. T. (2020). Evaluation of risk assessment of landfill emissions and their impacts on human health. *Arabian Journal of Geosciences*, 13, article 1185. https://doi.org/10.1007/s12517-020-06218-5.

Lin, X., Ma, Y., Chen, Z., Li, X., Lu, S., & Yan, J. (2020). Effect of different air pollution control devices on the gas/solid-phase distribution of PCDD/F in a full-scale municipal solid waste incinerator. *Environmental Pollution*, 265, article 114888. https://doi.org/10.1016/J.ENVPOL.2020.114888.

Manheim, D. C., Yeşiller, N., & Hanson, J. L. (2021). Gas emissions from municipal solid waste landfills: A comprehensive review and analysis of global data. *Journal of the Indian Institute of Science*, 101(4), 625–657. https://doi.org/10.1007/s41745-021-00234-4.

Marteinson, S. C., Lawrence, M. J., Taranu, Z. E., Kosziwka, K., Taylor, J. J., Green, A., Winegardner, A. K., Rytwinski, T., Reid, J. L., Dubetz, C., Leblanc, J., Galus, M. D., & Cooke, S. J. (2023). Increased use of sanitizers and disinfectants during the Covid-19 pandemic: Identification of antimicrobial chemicals and considerations for aquatic environmental contamination. *Environmental Reviews*, 31(1), 76–94. https://doi.org/10.1139/er-2022-0035.

Mohamed, B. A., Fattah, I. M. R., Yousaf, B., & Periyasamy, S. (2022). Effects of the Covid-19 pandemic on the environment, waste management, and energy sectors: A deeper look into the long-term impacts. *Environmental Science and Pollution Research*, 29, 46438–46457. https://doi.org/10.1007/s11356-022-20259-1.

Muir, D. C. G., & Galarneau, E. (2021). Polycyclic aromatic compounds (PACs) in the Canadian environment: Links to global change. *Environmental Pollution*, 273, article 116425. https://doi.org/10.1016/J.ENVPOL.2021.116425.

Nazari, L., Xu, C. C., & Ray, M. B. (2021). *Advanced and emerging technologies for resource recovery from wastes*. Springer. https://doi.org/10.1007/978-981-15-9267-6.

Ochs, C., Garrison, K., Saxena, P., Romme, K., & Sarkar, A. (2024). Contamination of aquatic ecosystems by persistent organic pollutants (POPs) originating from landfills in Canada and the United States: A rapid scoping review. *Science of the Total Environment*, 924, article 171490. https://doi.org/10.1016/J.SCITOTENV.2024.171490.

Parashar, N., Mahanty, B., & Hait, S. (2023). Microplastics as carriers of per- and polyfluoroalkyl substances (PFAS) in aquatic environment: interactions and ecotoxicological effects. *Water Emerging Contaminants & Nanoplastics*, 2(7), 1–21. https://doi.org/10.20517/wecn.2023.25.

Peula, F. J., María, Martín-Lara, Á., & Calero, M. (2023). Effect of Covid-19 pandemic on municipal solid waste generation: A case study in Granada city (Spain). *Journal of Material Cycles and Waste Management*, 25, 2543–2555. https://doi.org/10.1007/s10163-023-01671-2.

Sahle-Demessie, E., Mezgebe, B., Dietrich, J., Shan, Y., Harmon, S., & Lee, C. C. (2021). Material recovery from electronic waste using pyrolysis: Emissions measurements and risk assessment. *Journal of Environmental Chemical Engineering*, 9(1), article 104943. https://doi.org/10.1016/j.jece.2020.104943.

Schwartz, H., Marushka, L., Chan, H. M., Batal, M., Sadik, T., Ing, A., Fediuk, K., & Tikhonov, C. (2021). Pharmaceuticals in source waters of 95 First Nations in Canada. *Canadian Journal of Public Health*, 112, 133–153. https://doi.org/10.17269/s41997-021-00499-3.

Schwartz-Narbonne, H., Xia, C., Shalin, A., Whitehead, H. D., Yang, D., Peaslee, G. F., Wang, Z., Wu, Y., Peng, H., Blum, A., Venier, M., & Diamond, M. L. (2023). Per- and polyfluoroalkyl substances in Canadian fast food packaging. *Environmental Science and Technology Letters*, 10(4), 343–349. https://doi.org/10.1021/acs.estlett.2c00926.

Schwarz, A. E., Lensen, S. M. C., Langeveld, E., Parker, L. A., & Urbanus, J. H. (2023). Plastics in the global environment assessed through material flow analysis, degradation and environmental transportation. *Science of the Total Environment*, 875, article 162644. https://doi.org/10.1016/j.scitotenv.2023.162644.

Statistics Canada. (2011). *Households and the environment catalogue no. 11-526-X*. Minister of Industry, Government of Canada. www150.statcan.gc.ca/n1/en/pub/11-526-x/11-526-x2013001-eng.pdf?st=AHBuX3MA.

Statistics Canada. (2023). *Table 38-10-0032-01 Disposal of waste, by source*. Government of Canada. https://doi.org/10.25318/3810003201-eng.

Sühring, R., Baak, J. E., Letcher, R. J., Braune, B. M., de Silva, A., Dey, C., Fernie, K., Lu, Z., Mallory, M. L., Avery-Gomm, S., & Provencher, J. F. (2022). Co-contaminants of microplastics in two seabird species from the Canadian Arctic. *Environmental Science and Ecotechnology*, 12, article 100189. https://doi.org/10.1016/J.ESE.2022.100189.

Tripathi, V., & Tripathi, P. (2017). Antibiotic resistance genes: An emerging environmental pollutant. In K. K. Kesari (Ed.), *Perspectives in environmental toxicology* (pp. 183–201). Springer. https://doi.org/10.1007/978-3-319-46248-6_9.

Wang, F., Xiang, L., Sze-Yin Leung, K., Elsner, M., Zhang, Y., Guo, Y., Pan, B., Sun, H., An, T., Ying, G., Brooks, B. W., Hou, D., Helbling, D. E., Sun, J., Qiu, H., Vogel, T. M., Zhang,

W., Gao, Y., Simpson, M. J., … Tiedje, J. M. (2024). Emerging contaminants: A One Health perspective. *Innovation*, 5(4), article 100612. https://doi.org/10.1016/j.xinn.2024.100612.

Weichenthal, S., Van Rijswijk, D., Kulka, R., You, H., Van Ryswyk, K., Willey, J., Dugandzic, R., Sutcliffe, R., Moulton, J., Baike, M., White, L., Charland, J. P., & Jessiman, B. (2015). The impact of a landfill fire on ambient air quality in the north: A case study in Iqaluit, Canada. *Environmental Research*, 142, 46–50. https://doi.org/10.1016/J.ENVRES.2015.06.018.

Yáñez, J. A., Chung, S. A., Román, B. R., Hernández-Yépez, P. J., Garcia-Solorzano, F. O., Del-Aguila-Arcentales, S., Inga-Berrospi, F., Mejia, C. R., & Alvarez-Risco, A. (2021). Prescription, over-the-counter (OTC), herbal, and other treatments and preventive uses for Covid-19. In M. H. Dehghani, R. R. Karri, & S. Roy (Eds.), *Environmental and health management of novel coronavirus disease (Covid-19)* (pp. 379–416). Academic Press.

Yurtsever, M. (2019). Glitters as a source of primary microplastics: An approach to environmental responsibility and ethics. *Journal of Agricultural and Environmental Ethics*, 32(3), 459–478. https://doi.org/10.1007/s10806-019-09785-0.

Zhang, J., Yu, X., Wang, J., Sui, Q., & Zhao, W. (2023). Impacts of garbage classification and disposal on the occurrence of pharmaceutical and personal care products in municipal solid waste leachates: A case study in Shanghai. *Science of the Total Environment*, 874, article 162467. https://doi.org/10.1016/J.SCITOTENV.2023.162467.

16 Wind turbine end-of-life waste management in Nova Scotia, Canada

Mercy D. Fiamavle and Tony R. Walker

Introduction

Wind power is recognized globally as a sustainable and cost-effective energy source, offering significant environmental benefits over traditional fossil fuels (Macquart et al., 2022). Wind turbines (WTs) convert kinetic energy into electricity, contributing to local energy supplies and reduced greenhouse gas (GHG) emissions (Parisé & Walker, 2017). The imperative to combat climate change and enhance energy security has intensified adoption of renewable energy sources like wind power (Cherrington et al., 2012). Achieving the ambitious Paris Agreement targets requires substantial reductions in GHG emissions, emphasizing the critical role of wind power. According to the International Energy Agency (IEA), 77.6 gigawatts (GW) of new wind power capacity was installed globally in 2022, highlighting the accelerating shift towards renewable energy adoption (GWEC, 2023).

In Canada, over 80% of current GHG emissions originate from energy generation across various sectors, underscoring the urgency for transitioning to renewable energy to achieve net-zero emissions by 2050 (CER, 2023a). This transition will reshape material demands and waste generation in the energy sector, necessitating a focus on recovering value from waste and addressing end-of-life (EOL) considerations. Projections indicate significant WT waste volumes, estimated at 831,000 tons (t) by 2030 and 4,500,000 t by 2050 (Cairns & McCarney, 2021). While materials like concrete, iron, and steel in turbine components are highly recyclable (WindEurope, 2020), challenges remain in recycling rotor blades due to their size and composition, with blade waste projected to reach 25,000 t by 2030 and 135,000 t by 2050 (Cairns & McCarney, 2021).

This chapter discusses the urgent need for a robust WT EOL management in Nova Scotia (NS), draws insights from global best practices, identifies gaps in current frameworks, and assesses environmental impacts. It also forecasts WT waste volumes and decommissioning timelines for NS, and proposes emerging solutions and strategies used by global leaders in EOL management.

DOI: 10.4324/9781003534129-22

Wind energy in NS

Located in eastern Canada, NS aims to reach net-zero GHG emission goals (23% by 2030) set by the federal government (Giacosa & Walker, 2022). NS has legislated the *Environmental Goals and Climate Change Reduction Act* and the Renewable Energy Regulations made under the *Electricity Act* (CCI, 2020). The province is actively progressing towards creating more environmentally sustainable energy such as hydro, solar, wind, and tidal for the future and by so doing looking to phase out coal and striving to achieve 80% renewable energy target by 2030 (Giacosa & Walker, 2022). NS is home to over 300 commercial onshore WTs making the province one of Canada's leaders in wind energy. As of 2022, NS installed 603 MW of wind and 0.37 MW of solar on its grid (14% of total generation).

In NS, wind energy projects of atleast 2 MW must undergo a provincial Environmental Assessment (EA) under the Environmental Assessment Regulations, N.S. Reg. 328/2022, overseen by the Department of Environment. *A Proponent's Guide to Wind Power Projects (APGWPP): Guide for Preparing an Environmental Assessment (EA) Registration Document* offers guidance for proponents (CCI, 2020). NS has committed to reducing its carbon footprint, targeting 5 GW of offshore wind energy by 2030 with substantial federal and provincial investments. Investments include CAD 125 Million (M) for new wind projects and CAD 11.5 M for clean power initiatives. Wind energy initiatives like the Higgins Mountain wind farm is set to launch in 2024, aiming to generate 372 MW annually by 2025, contributing significantly to NS's energy needs (Government of Canada, 2023). Beyond energy benefits, the wind sector fosters economic growth, job creation, and infrastructure development (US Department of Energy, 2023). However, there is a need to address new challenges in WT EOL management and reduce associated carbon footprints (European Commission, 2023). WT rotor blades can measure up to 80 m, weigh 35 t, and are primarily made from glass or carbon fiber-reinforced polymer (FRP), featuring a core of balsa wood or plastic foam, supported by shear webs (ACP Wind Blade Working Group, 2023). Typically, a blade's material composition includes 60% reinforced fibers (mainly glass), 23% thermosetting polymer resins and adhesives, 9% core materials like balsa wood or thermoplastic foams, and 8% metals such as copper and steel (Fingersh et al., 2006). Table 16.1 lists the materials used in WTs.

WTs EOL management in NS

Decommissioning wind farms involves dismantling WT and restoring the site back to its original state (Ortegon et al., 2013). Decommissioning responsibilities for onshore wind farms are governed by provincial and municipal authorities. The NS Environmental Assessment Regulations, N.S. Reg. (NS EAR) 328/2022, and APGWPP provide the regulatory requirements for preparing the EA for wind farm project construction, operation, and decommissioning phases. Commercial wind farms, with at least a 2 MW production rating are classified as Class I undertakings (Environmental Assessment Regulations, 2022) and are subject to provincial

Table 16.1 Percentage of materials used in WT components

Component/material (% by weight)	*Steel*	*Copper*	*Aluminum*	*Glass/ reinforced plastic*	*Prestressed concrete*
Rotor					
Hub	100				
Blades	5	*	*	95	
Nacelle	80	14	3–4	1	
Gearbox	98	2	2		
Generator	65	35			
Frame, machinery, and shell	85	4	9	3	
Tower	98				2

Source: Ancona and McVeigh (2001).

Note: *According to Fingersh et al. (2006), copper-steel and other metal composites in blades comprise roughly 8%, but most of this comprises of steel with copper and other metals used in smaller proportions.

regulations. Additionally, municipalities like Cumberland and Colchester have specific by-laws: Cumberland's Municipal Planning Strategy & Land Use By-law and Colchester's Chapter 56 Wind Turbine Development Bylaw (Municipality of Cumberland, 2022; Municipality of Colchester, n.d.).

Although wind energy is a sustainable energy source, environmental effects are evident throughout the entire life cycle of wind projects. Impacts are particularly evident during manufacturing, transportation, installation, maintenance, and, notably, at the EOL stage (Mello et al., 2002). Recycling WT components is challenging due to complex composite materials and thermoset resins, combined with the lack of suitable recycling methods (EWEA, n.d.). By 2050, global blade waste is projected to reach 43 M tonnes(t), with 2 M t added annually (Liu & Barlow, 2017). The wind energy sector is set to become Europe's primary source of composite waste by 2050 (ETIP Wind, 2020). EOL management of turbine blades is crucial, with landfilling being the most economical but least preferred option due to the lack of material recovery (D'Souza et al., 2001). Alternatives include repowering, upgrading turbines to extend lifespans to 40–50 years, refurbishing components for efficiency, and decommissioning with full removal and site restoration under "polluter pays" and Extended Producer Responsibility (EPR) principles (Diggle & Walker, 2020).

Methods

Relevant literature was reviewed by examining peer-reviewed and gray literature to understand management challenges, impacts, and sustainable solutions within the wind energy industry across various jurisdictions. Using databases such as "Scopus", "ProQuest", "Web of Science", "ResearchGate", ScienceDirect,

SpringerLink, Factiva, and "Google Scholar", keywords including "Nova Scotia," "wind turbine," "end-of-life," "decommissioning," "innovations and methods for EOL management of wind turbine blade waste," "wind turbine waste management policies", and "environmental impacts of wind turbine waste" were used to help refine the scope of the search. Relevant data were sorted based on these keywords and research questions. Additionally, gray literature, including government documents, agency technical reports, news articles, and blog posts were reviewed. A review of NS EAR 328/2022 and APGWPP were conducted to assess frameworks for WT EOL management. A generic search using terms like "decommissioning" and "turbine waste" found limited peer-reviewed literature, mostly focused on Life Cycle Assessment (LCA), which was outside the scope of the study. Gray literature from government sources were therefore retrieved to compensate for the lack of peer-reviewed work on WT waste management in NS.

A list of all the wind farm projects in NS was retrieved from the Canadian WT database (Natural Resources Canada, 2021). A total of 310 turbines have been installed in NS as of 2021. The following information was collected for wind projects:

- Project name
- Manufacturer
- Technical data: model, turbine rated capacity (kW), rotor diameter, and hub height
- Date of commissioning.

A high-level comparison of EAs for wind farms in NS was conducted to determine if WT blade waste management was addressed. The analysis focused on the 12 largest wind farms by turbine count. EA documents were searched on the Government of NS website using keywords such as "decommissioning," "end-of-life," and "turbine waste" (GNS, 2019). Blade material estimates were based on manufacturer-reported blade weights, assumed to be glass fiber reinforced plastic (GFRP) with ±60% glass fiber and ±40% polymer matrix (WindEurope, 2020). Due to data gaps, an average blade weight of ±35 t was used, multiplied by three for the number of blades per turbine (Jensen et al., 2020). Waste blade volumes were calculated using an average fiberglass density of ±1,880 kg/m^3 and typical blade composition (Ancona & McVeigh, 2001; JPS Composite Materials, 2021). This method provided reasonable estimates of waste material by dividing the mass by the composite material's density (Heng et al., 2021).

Limitations

Limitations of this study include the availability of WT EOL management practices and evolving technology, regulations, and industry practices. Despite these, the study will inform future research on policy development and best practices to ensure long-term environmental sustainability for wind energy projects in NS.

Results and discussion

A total of 310 turbines were manufactured by five main companies including Enercon (40%), General Electric (GE; 24%), Vestas (16%), Acciona (11%) Suzlon (5%), and Others (4%; Figure 16.1).

The review of the NS EAR 328/2022 and APGWPP showed there were no regulatory requirements or guidelines for EOL management of WT waste. However, the guidance document requires that the proponent provide goals and objectives for decommissioning but does not require the proponent to submit plans for the EOL management of WT. Table 16.2 shows the results of the analysis of the 12 EA documents of the windfarms. Only Amherst I Wind Farm stated management plans for WT waste. Three windfarms stated that "the waste will be disposed of in a manner that meets all applicable regulatory requirements". The rest made no specific management plans for EOL WT waste.

Development of wind energy in NS is predominantly young, however, the situation will change in the coming years as the WTs reach their typical 20–25-year design life and intended life expectancy (Ortegon et al., 2013). The projected growth in onshore and offshore wind development will lead to the establishment of numerous WTs. The estimated number of WTs that will reach their planned service life in NS is shown in Figure 16.2. Between 2025 and 2030, 178 WTs will reach EOL. From 2030 to 2040, 132 WTs will be facing these EOL scenarios in NS. An

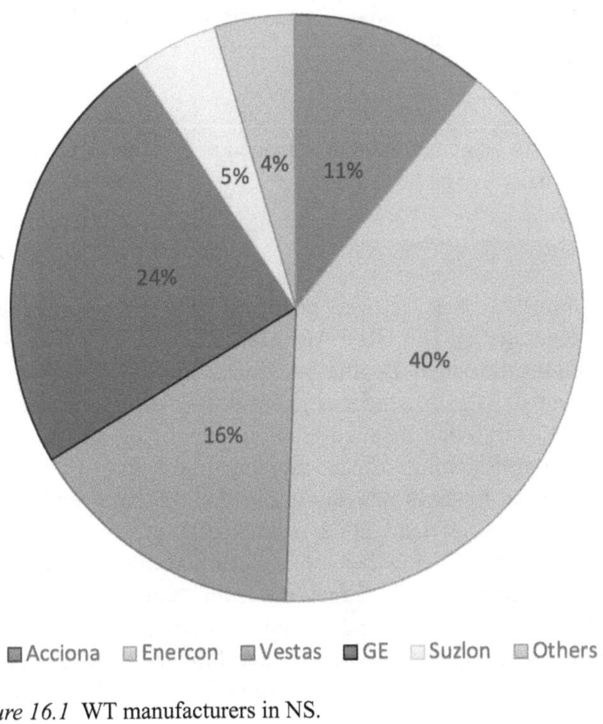

Figure 16.1 WT manufacturers in NS.

Table 16.2 EOL plans in EA documents for 12 largest windfarm projects in NS

No.	*Project name*	*No. of turbines*	*Decommissioning plans made in EA*
1	Dalhousie Mountain	34	No EOL management commitments
2	South Canoe	34	WTs will be dismantled and removed from the site
3	Glen Dhu	27	No EOL management commitments
4	Nuttby	22	Disposed of in a manner that meets all applicable regulatory requirements
5	Digby Neck	20	No EOL management commitments
6	Pubnico Point	17	Disposal will be undertaken in accordance with the regulatory regime in place at the time
7	Amherst I Wind Farm	15	Waste may be sold to recycling facilities, or crushed and deposited at the landfill
8	Point Tupper	12	Dismantled and disposed of in a manner that meets all regulatory requirements
9	Ellershouse	10	No EOL management commitments
10	Lingan	8	No EOL management commitments
11	Sable	6	WTs will be dismantled and removed from the site, A decommissioning plan will be submitted
12	Chebucto Pockwock	5	NSECC will be provided with decommissioning plan

Source: Adopted from DNRR (2021).

estimate of waste volumes to be generated and managed at each decommissioning date is shown in Figure 16.3.

Legal framework

In NS, establishment of onshore wind farms is governed by both provincial legislation and municipal regulations. The NS EAR 328/2022 and APGWPP serve as the primary regulations provided by the province, outlining requirements for the EOL management of WT in wind energy projects. The guidance document requires proponents to provide detailed plans for removing structures and restoring sites post-decommissioning. Some municipalities, like Cumberland and Colchester, have enacted by-laws mandating decommissioning bonds from proponents (Municipality of Cumberland, 2022; Municipality of Colchester, n.d.). These bonds cover the costs of removing decommissioned turbines and structures; however, there are no regulations or guidelines governing safe dismantling and waste management. In Europe, stricter landfill and incineration regulations are emerging, with Germany, Austria, the Netherlands, and Finland leading efforts to prohibit landfilling, reflecting a shift toward environmentally conscious practices (Bhatti, 2021).

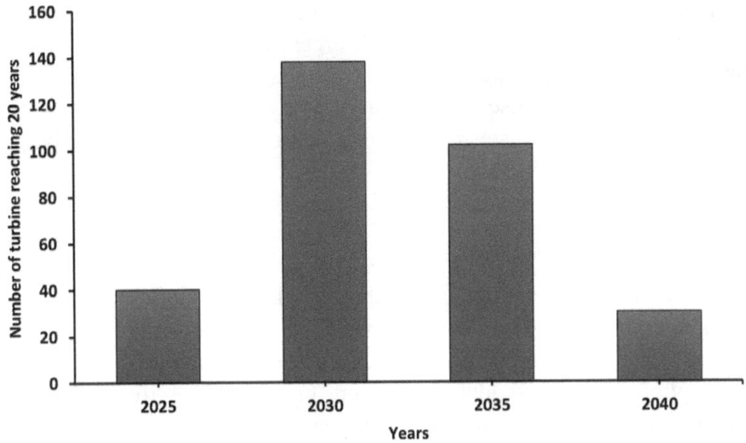

Figure 16.2 Number of WT reaching EOL by 2040 in NS.

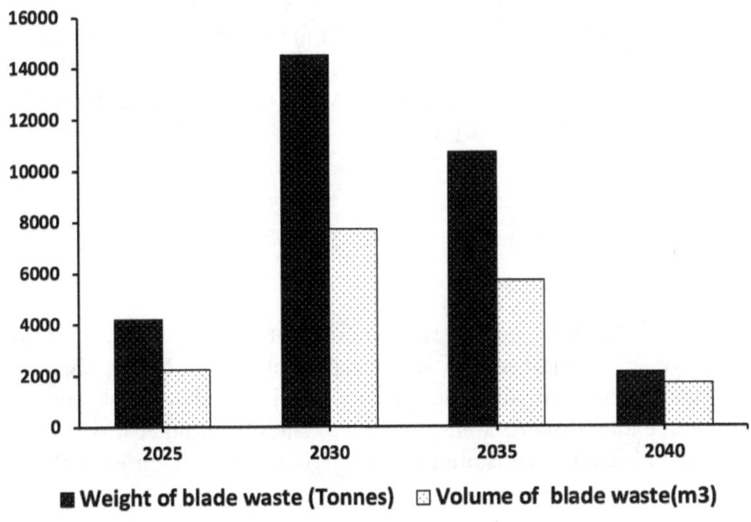

Figure 16.3 Estimated weight and volumes of WT blade waste.

Estimated weight and volumes of wind blade composite material

Findings suggest that approximately ±31,500 t of wind blade composite material will require management by 2040 in NS. This data, coupled with estimated decommissioning dates, offers valuable insights for policymakers, the wind industry, and local governments to develop waste management strategies. The results highlight a significant challenge for NS's wind industry due to the potential large volume of about ±17,298 t of blade waste material. The decommissioning of around 20 WT in 2025 highlights the need for urgent action. Estimating blade waste dates can aid in developing repurposing strategies, supporting a circular economy by extending blade material use and delaying waste entry. This proactive approach ensures local, sustainable waste management and avoids carbon-intensive transportation (Delaney et al., 2021).

EOL environmental and health impacts

Impacts can be categorized into various classifications including CO_2 emissions, climate change, cumulative energy demand, abiotic depletion, acidification, stratospheric ozone depletion, human toxicity, particulate matter formation, dust, ecotoxicity, photochemical oxidation, nutrient enrichment, eutrophication, solid waste generation, land use, land transformation, human health endpoint, natural environment endpoint, natural resources endpoint, single score endpoint, nontoxic emissions, and toxic emissions (Arvesen et al., 2012). With an expected ±31,500 t of GFRP composite material from WT in NS by 2040, landfill disposal emerges as the least desirable waste management option within the waste hierarchy, due to sustainability and practicality concerns. The bulky nature of composite blades leads to a high volume-to-weight ratio, making them unsuitable for conventional landfilling, which is designed for compact waste and limited space. Their large size and volume create challenges for efficient waste management and compaction. While landfill disposal is currently permitted in NS, it should be a least desired resort (WindEurope, 2020).

Management challenges of WT waste

Waste preparation and transportation

WT waste is initially sectioned into smaller sizes for transport to waste treatment facilities. However, the large size of WT occupies significant land space and may reduce the capacity originally intended for land use. Located in remote areas, wind farms necessitate long-distance transportation to landfills or other waste treatment sites, resulting in fossil fuel consumption and GHG emissions (Jensen et al., 2020).

Landfilling

WT blade materials are inert in landfills and do not immediately release GHGs like CH_4 and CO_2 but their large volume consumes significant landfill space. The plastic components can break down into microplastics and nanoplastics, causing groundwater and ocean pollution. The extensive land required for disposal impacts conservation efforts, leading some Ontario municipalities to reject decommissioned blades to address these issues. (Heng et al., 2021; Duan et al., 2021; He et al., 2019).

Incineration

Incinerating EOL WT waste is problematic due to the high glass fiber content, which releases harmful substances like halogenated additives, polyvinyl chloride fumes, furans, dioxins, and polychlorinated biphenyls. The low thermal efficiency of fiberglass (35%) and hazardous fly ashes pose risks, including health issues such as soft tissue sarcomas and reduced lung function among incinerator workers (Rathore & Panwar, 2023; Alabi et al., 2019; Papadakis et al., 2010; Zambon et al., 2007).

Mechanical recycling

Mechanical recycling of waste WT involves shredding glass and carbon FRP into 50–100 mm pieces, further reducing size with a hammer mill, and separating materials into fiber and powder-rich fractions. This process can produce fine dust that may be released into the air, potentially causing skin, eye, nose, and throat irritation, and with prolonged exposure, could lead to asthma or bronchitis (Zhang et al., 2020).

Sustainable alternative EOL pathways

Given the rapid growth of wind energy globally, alternative EOL solutions for WT blades are imperative. Past explorations have considered mechanical and thermal recycling options (Khalid et al., 2023). Although landfilling is the most economical management option for wind blades, it is unsustainable, prompting the search for more sustainable alternatives (ACP Wind Blade Working Group, 2023). Various jurisdictions prioritize sustainable waste management pathways following the waste hierarchy: Prevention (blade life extension), Reuse (second-hand market), Repurpose (creating new products from blades), Recycle (downcycling into raw materials), Recovery (converting nonrecyclable waste into usable energy), and Disposal (landfilling). EOL options must align with the Canadian Waste Hierarchy to uphold the sustainability of wind energy (Government of Canada, 2021).

While these options exist, not all are financially viable or scalable, though they hold promise for future EOL solutions for WT blades. Some governments have addressed regulatory challenges; for instance, some states in the United States such as Washington are addressing regulatory challenges by proposing bans on landfill disposal or requiring manufacturers to reclaim a percentage of blade waste through take-back programs (ACP Wind Blade Working Group, 2023). Since 2014, Scottish Planning Policy in the UK mandates perpetual suitability of wind farm sites and extends lease durations to 50–60 years to enhance component durability (Hutchings et al., 2023). Efforts to ban landfilling of WT blades are underway to promote alternative methods such as repowering, lifetime extension (LTE), refurbishing, reusing, repurposing, and recycling (Woo & Whale, 2022). By 2022, Germany, Netherlands,

Austria, and Finland have implemented bans on landfilling turbine blades, with more European countries expected to follow suit by 2025 (Bhatti, 2021).

Repowering or lifetime extension

Approximately 178 WTs in NS will require decisions on LTE, repowering, or decommissioning between 2025 and 2030, with nearly 132 WTs facing these scenarios from 2030 to 2040. In Europe, repowering has been undertaken at 170 wind farms, led by the Netherlands, and followed by Germany (WindEurope, 2022). Most wind farms approaching EOL currently opt for LTE, as legislative frameworks for full repowering are often lacking (Wind Energy Technologies Office, 2021). Full repowering involves decommissioning entire wind farms, including towers and foundations, and replacing original turbines with more powerful units in a revised layout (Serri et al., 2018). Partial repowering replaces components like drivetrains, gearboxes, and rotors while retaining foundations and cables. This approach can potentially triple energy output, extend turbine lifespan on secured land, and leverage tax incentives (Wiser et al., 2022). Repowering will enable NS to achieve its 80% renewable energy target by 2030 without losing wind energy capacity through decommissioning, reducing waste, and providing economic incentives (IWEA, 2019). Repowering remains a key strategy for maintaining wind energy capacity but few wind farms in Canada have been decommissioned or repowered due to the relatively young age of fleets and good maintenance practices encouraged by the Canada Renewable Energy Association (CanRea, n.d.).

Repurposing

WT blades unfit for reuse in their original capacity can be repurposed for innovative applications. Research by the Rewind Network has turned decommissioned blades into "BladeBridges" in Ireland (Bank et al., 2017). Other applications explored include bus shelters, barriers, street furniture, and telecommunication towers (Gallapher, 2023). In the Netherlands, Superuse Studios has created housing structures such as roofs, doors, stations, playgrounds, elevated foundations, and window frames (Rathore & Panwar, 2023). Blades can also be repurposed into bridges, and housing projects, especially in areas prone to harsh environmental conditions (Suhail et al., 2019). In NS, the use of WT blade composites for wall panels, roofing, or structural supports is ideal for harsh weather. These materials' strength and resistance to environmental stresses make them well suited for enduring severe conditions like hurricanes, rainstorms, strong winds, and temperature fluctuations, enhancing home resilience.

Recycling

Recycling WT blades provides a second life for the material, reduces the demand for virgin resources, and recovers valuable constituents. Mechanical recycling involves crushing and grinding GRFC materials, thermal recycling uses heat for fiber recovery, and chemical recycling employs dissolution agents to reclaim fibers. (Rathore & Panwar, 2023; Oliveux et al., 2015). In Germany, Geocycle burns thousands of tons of old blades in cement kilns, emitting less carbon dioxide than fossil fuels (Neslen, 2021). Goldwind, a Chinese wind company, aims to recycle all its blades by 2040, creating 3D-printed products like flowerpots and garden chairs utilizing ground composites. Vestas, a leading WT manufacturer, targets zero-waste WTs by 2040 through initiatives like DecomBlades for recycling existing blades and CETEC for developing new resin technologies within a circular economy framework (Vestas, 2023). A circular economy in blade recycling depends on the effective management of large volumes of blade waste and blade material identification (Jensen et al., 2020).

Recommendations

Develop EOL framework

Legislation serves as a powerful tool for promoting the adoption of sustainable practices (Cherrington et al., 2012). Although different strategies are possible, legislation can focus on WT installations, manufacturers, or generated waste to reinforce reuse, recycling, and recovering processes (Beauson & Brøndsted, 2016). Current policies that impede the progress of the wind energy sector toward a circular economy must be updated. For example, China's 2023 guidelines from the National Development and Reform Commission aim to standardize wind facility decommissioning and recycling, creating a "one-stop" service model covering the entire lifecycle from dismantling to reuse. A framework for planning and regulating the decommissioning of wind energy facilities must be established by stakeholders including Environment and Climate Change Canada, NS Department of Environment and Climate Change (NSECC), nongovernmental organizations, wind energy companies, WT manufacturers, industry associations like the Canadian Wind Energy Association (CanWEA), academic and research institutions, recycling and waste management experts and the indigenous communities. Additionally, circular economy strategies and policies should be implemented to support the life of WTs and make it easier to recover value and close material loops, while diverting waste from landfills (Cairns & McCarney, 2021).

Implement EPR program

EPR is a policy approach under which producers are given a significant responsibility either financial and/or physical for treatment or disposal of post-consumer products (Diggle & Walker, 2020). EPR policies shift the responsibility for waste management from the consumer or public authorities to product manufacturers. EPR programs encourage manufacturers to design for recycling to close material loops, encourage the use of waste material as raw material and invest in innovative recycling solutions leading to a significant reduction in landfill waste. Currently, EPR principles have not been widely to the wind energy sector as it is relatively new compared to other industries such as the automotive, electronic, and electrical equipment where EPR frameworks are already well-established (Cherrington et al., 2012), however, this narrative can be changed by the introduction of EPR measures by policymakers for the wind energy sector to ensure dedicated and sufficient funding for the treatment of EOL of WTs (Hutchings et al., 2023). Federal and provincial governments can spearhead initiatives to better regulate and promote the reuse of WT components by setting targets for refurbished parts measurable in terms of federal carbon savings. To support this, governments should establish clear regulations, create financial incentives for compliance, and mandate regular progress reporting. Monitoring systems should be implemented to track compliance, foster industry collaboration and develop best practices. Lastly, the government should establish enforcement mechanisms that include penalties for noncompliance and incentives for exceeding targets.

Stakeholder collaboration

Stakeholder collaboration involves engaging government, industry, local communities, and environmental organizations. This inclusive approach ensures a comprehensive strategy that considers diverse perspectives and expertise. Policymakers should promote data sharing among key stakeholders for WT technology. Research institutions and organizations such as CanWEA should collaborate with major WT manufacturers such as Vestas, Enercon, and GE to drive innovation and identify best practices for development and adoption of advanced technologies and methodologies for treating waste WT blade composites (Arias, 2016). Collaboration with industry stakeholders such as manufacturing, power generation, recycling, concrete and housing sectors can open avenues for economic opportunities, including job creation and development of a skilled workforce, ultimately contributing to regional economic growth.

Develop a decommissioning database

Existing WT, whether repowered or completely decommissioned, will need to be dismantled and hence planning for waste materials and blade removal is essential. To evaluate potential options for EOL of wind blades, it is crucial to understand the sources and time of production of the waste blade material. An organization should be established by NSECC, to oversee the development of a decommissioning database to centralize the information such as manufacturer, type, number, quantity, decommissioning dates, and location of the WT for waste feedstock planning. This information enables a detailed waste stream analysis during decommissioning, optimizing resource recovery and providing insights into the recyclability of components for forecasting future waste volumes.

Construct turbine blade waste facilities

The development of specialized waste facilities offers a regulated system for the disposal of turbine blade waste. These dedicated facilities can serve as repositories for waste, offering opportunities for research, technology, and innovation to explore alternative sustainable waste treatment pathways that ensure circularity and further enhance the environmental sustainability of wind energy. Additionally, the potential environmental and health impacts associated with prevalent cost-effective disposal methods like landfilling or incineration can be reduced. By investing in these waste storage facilities, NS can demonstrate a strong commitment to wind energy sustainability. This ensures that EOL considerations are integrated into the turbine lifecycle, reducing the ecological footprint associated with turbine decommissioning.

Challenges and solutions with implementing recommendations

The implementation of these recommendations may encounter challenges, including delayed policy updates, high initial costs, and resistance from stakeholders, particularly manufacturers. Industry pushback on EPR programs, significant investment for recycling infrastructures, and local community opposition to waste management facilities due to environmental and health concerns. These challenges can be addressed by involving all relevant stakeholders (manufacturers, recycling companies, local communities, and government agencies) in the policy-making process to ensure support and address concerns. Pilot projects can test and refine frameworks before full implementation. Financial incentives or subsidies can help offset the initial costs; an example of this is the United States (US) tax credits for repowering wind turbines (WTs). Brookfield Renewable Partners for example is investing $600 M to upgrade its 845 MW Shepherds Flat wind project in Oregon, US. Government funding and partnerships with academic institutions are vital for establishing and maintaining databases, as well as leveraging their expertise and resources to advance sustainability goals.

Conclusion

As wind energy demand grows, managing EOL waste from WTs becomes imperative for reducing environmental harm and promoting sustainability. In NS, WTs will reach their EOL over the next few years, with an estimated 31,500–35,000 t of blade material expected by 2040. Without decommissioning plans, NS must address EOL management and implement strategies as waste volumes increase. This chapter proposes five solutions: develop an EOL framework, implement EPR programs, foster stakeholder collaboration, create a decommissioning database, and build waste storage facilities. Future studies could explore blade specifics and improve data access. NS and Canada should learn from practices used in the US and Europe to develop effective policies and regulations.

References

ACP Wind Blade Working Group. (2023). *Decommissioned wind turbine blade management strategies*. https://cleanpower.org/wp-content/uploads/2023/01/ACP_BladeRecycling_WhitePaper_230130.pdf.

Alabi, O. A., Ologbonjaye, K. I., Awosolu, O., & Alalade, O. E. (2019). Public and environmental health effects of plastic wastes disposal: A review. *Journal of Toxicology and Risk Assessment*, 5, 021. https://doi.org/10.23937/2572-4061.1510021.

Ancona, D., & McVeigh, J. (2001). *Wind turbine materials and manufacturing factsheet.* Princeton Energy Resources International.

Arias F. (2016). *Assessment of present/future decommissioned wind blade fiber-reinforced composite material in the United States.* Department of Civil Engineering. www.researchgate.net/publication/338430512_Assessment_of_PresentFuture_Decommissioned_Wind_Blade_Fiber-Reinforced_Composite_Material_in_the_United_States.

Arvesen, A., & Hertwich E. G. (2012). Assessing the life cycle environmental impacts of wind power: A review of present knowledge and research needs. *Renewable and Sustainable Energy Reviews*, 16, 8, 5994–6006. https://doi.org/10.1016/j.rser.2012.06.023.

Bank, L. C., Arias, F. R., Gentry, R., et al. (2017). Reusing composite materials from decommissioned wind turbine blades. In *17th International Conference on Non-conventional Materials and Technologies (NOCMAT 2017)* held in Yucatán, Mexico.

Beauson, J., & Brøndsted, P. (2016). Wind turbine blades: An end-of-life perspective. In *Mare-Wint: New materials and reliability in offshore wind turbine technology* (pp. 421–432). Cham, Springer. https://doi.org/10.1007/978-3-319-39095-6_23.

Bhatti, J. (2021). *Europe bans disposal of decommissioned wind turbine blades in landfills: A step towards life cycle sustainability.* www.downtoearth.org.in/blog/energy/europe-bans-disposal-of-decommissioned-wind-turbine-blades-in-landfills-a-step-towards-life-cycle-sustainability-77835.

Cairns, S., & McCarney, G. (2021). *Solar and wind energy in Canada: Value recovery and end-of-life considerations.* Smart Prosperity Institute. https://institute.smartprosperity.ca/sites/default/files/Policy%20Brief%20-%202021%20-%20Solar%20Wind%20-%20October20.pdf.

Canada Energy Regulator. (2023a). *Canada Energy Regulator projects a major transformation to Canada's energy system in a net-zero world.* www.cer-rec.gc.ca/en/about/news-room/news-releases/2023/canada-energy-regulator-projects-major-transformation-canada-energy-system-net-zero-world.html.

Canadian Climate Institute. (2020). *Clean growth in Nova Scotia.* https://climateinstitute.ca/publications/clean-growth-in-nova-scotia.

Cherrington, R., Goodship, V., Meredith, J., et al. (2012). Producer responsibility: Defining the incentive for recycling composite wind turbine blades in Europe. *Energy Policy*, 47, 13–21. https://doi.org/10.1016/j.enpol.2012.03.076.

Delaney, E. L., McKinley, J. M., Megarry, W., et al. (2021). An integrated geospatial approach for repurposing wind turbine blades. *Resources, Conservation & Recycling*, 170, 105601.

Department of Natural Resources and Renewables. (2021). *Wind energy in Nova Scotia.* https://energy.novascotia.ca/renewables/wind-energy.

Diggle, A., & Walker, T. R. (2020). Implementation of harmonized extended producer responsibility strategies to incentivize recovery of single-use plastic packaging waste in Canada. *Waste Management*, 110, 20–23. https://doi.org/10.1016/j.wasman.2020.05.013 Get rights and content.

D'Souza, N., Gbegbaje-Das, E., & Shonfield, P. (2001). *Raw metal needs and supply risks for the development of wind energy in Germany until 2050.* PE Northwest Europe ApS. https://doi.org/10.1016/j.jclepro.2019.02.223

Duan, J., Bolan, N., Li, Y., et al. (2021). Weathering of microplastics and interaction with other coexisting constituents in terrestrial and aquatic environments. *Water Research*, 196, 117011. https://doi.org/10.1016/j.watres.2021.117011.

Environmental Assessment Regulations, N.S. Reg. 328/2022.

Environmental Goals and Climate Change Reduction Act. 2021, c. 20, s. 1.

ETIPWind. (2020). *How wind is going circular blade recycling.* https://etipwind.eu/files/reports/ETIPWind-How-wind-is-going-circular-blade-recycling.pdf.

European Commission. (2023). *More circular, less carbon: chemical recycling holds promise for wind turbine blade waste.* European Commission Directorate-General for Environment. https://environment.ec.europa.eu/news/more-circular-less-carbon-chemical-recycling-holds-promise-wind-turbine-blade-waste-2023-10-19_en.

European Wind Energy Association. (n.d). *Research note outline on recycling wind turbines blades.* www.ewea.org/fileadmin/files/our-activities/policy-issues/environment/research_note_recycling_WT_blades.pdf.

Fingersh, L., Hand, M., Laxson, A., et al. (2006). Wind turbine design cost and scaling model. *No. NREL/TP-500-40566.* National Renewable Energy Lab. www.nrel.gov/docs/fy07osti/40566.pdf.

Gallapher, E. (2023). *Researchers transform landfilled wind turbine blades into footbridges.* www.qub.ac.uk/News/Allnews/2023/wind-turbines-turned-into-footbridges.html

Giacosa, G., & Walker, T. R. (2022). A policy perspective on Nova Scotia's plans to reduce dependency on fossil fuels for electricity generation and improve air quality. *Cleaner Production Letters*, 3, 100017. https://doi.org/10.1016/j.clpl.2022.100017.

Global Wind Energy Council. (2023). *Global wind report 2023.* https://gwec.net/wp-content/uploads/2023/04/GWEC-2023_interactive.pdf.

Government of Canada. (2021). *Reducing municipal solid waste* www.canada.ca/en/services/environment/weather/climatechange/climate-plan/clean-electricity/overview-nova-scotia.html

Government of Canada. (2023). *Nova Scotia: Clean electricity snapshot.* www.canada.ca/en/services/environment/weather/climatechange/climate-plan/clean-electricity/overview-nova-scotia.html

Government of Nova Scotia. (2019). *Projects environmental assessment Nova Scotia environment.* https://novascotia.ca/nse/ea/projects.asp.

He, P., Chen, L., Shao, L., et al. (2019). Municipal solid waste (MSW) landfill: A source of microplastics? -Evidence of microplastics in landfill leachate. *Water Research*, 159, 38–45. https://doi.org/10.1016/j.watres.2019.04.060.

Heng, H., Meng, F., & McKechnie, J. (2021). Wind turbine blade wastes and the environmental impacts in Canada. *Waste Management*, 133, 59–70. https://doi.org/10.1016/j.wasman.2021.07.032.

Hutchings, T., Joan, M. N., Tushita, R., et al. (2023). *For a circular energy transition action plan for industry, policymakers and investors.* https://circulareconomy.europa.eu/platform/sites/default/files/2023-06/Circular-energy-transition.pdf.

International Wind Association. (2019). *More power to you: A guide to repowering in Ireland.* www.iwea.com/images/files/repoweringes.pdf.

Jensen, P., Velenturf, A., & Purnell, P. (2020). Highlighting the need to embed circular economy in low carbon infrastructure decommissioning: The case of offshore wind. *Sustainable Production and Consumption*, 24, 266–280. https://doi.org/10.1016/j.spc.2020.07.012.

JPS Composite Materials. (2021). *E-glass & S-glass.* https://jpscm.com/products/e-glass-s-glass/

Khalid, M. Y., Arif, Z. U., Hossain, M., & Umer, R. (2023). Recycling of wind turbine blades through modern recycling technologies: A road to zero waste. *Renewable Energy Focus*, 44, 373–389.

Liu, P., & Barlow, C.Y. (2017). Wind turbine blade waste in 2050. *Waste Management*, 62, 229–240.

Macquart, T., Pirrera, A., & Weaver, P. (2022). *Promoting sustainability by solving wind turbine design challenges.* Bristol Composites Institute Blog. https://composites.blogs.bristol.ac.uk/2022/10/04/promoting-sustainability-by-solving-wind-turbine-design-challenges.

Mello, G., Ferreira Dias, M., & Robaina, M. (2022). Evaluation of the environmental impacts related to the wind farms end-of-life. *Energy Reports*, 8, 35–40. https://doi.org/10.1016/j.egyr.2022.01.024.

Municipality of Colchester. (n.d.). *Chapter 56: The Colchester wind turbine development by-law.* www.colchester.ca/4371-wind-turbine-development-by-law-chapter-56-september-2024/file

Municipality of Cumberland. (2022). *Plan Cumberland municipal planning strategy*. https://static1.squarespace.com/static/5845c0c6f5e231a284532a4a/t/6298bb6797dade5cd14c573d/1654176616543/Exerpt+from+Cumberland+MPS+-+Draft+for+PH.pdf.

Natural Resources Canada. (2021). *Canadian wind turbine database.* https://search.open.canada.ca/openmap/79fdad93-9025-49ad-ba16-c26d718cc070.

Oliveux, G., Dandy, L. O., & Leeke, G. A. (2015), Current status of recycling of fibre reinforced polymers: Review of technologies, reuse and resulting properties. *Progress in Materials Science*, 72, 61–99. https://doi.org/10.1016/j.pmatsci.2015.01.004.

Ortegon, K., Nies, L. F., & Sutherland, J. W. (2013). Preparing for end of service life of wind turbines. *Journal of Cleaner Production*, 39, 191–199. https://doi.org/10.1016/j.jclepro.2012.08.022.

Papadakis, N., Ramírez, C., & Reynolds, N. (2010). Designing composite wind turbine blades for disposal, recycling or reuse. In V. Goodship (Ed.), *Management, recycling and reuse of waste composites* (pp. 443–457). Elsevier. https://doi.org/10.1533/9781845697662.5.443.

Parisé, J., & Walker, T. R. (2017). Industrial wind turbine post-construction bird and bat monitoring: A policy framework for Canada. *Journal of Environmental Management*, 201, 252–259. https://doi.org/10.1016/j.jenvman.2017.06.052.

Rathore, N., & Panwar, N. L. (2023). Environmental impact and waste recycling technologies for modern wind turbines: An overview. *Waste Management & Research*, 41(4), 744–759. https://doi.org/10.1177/0734242X221135527.

Serri, L., et al. (2018). Wind energy plants repowering potential in Italy: Technical-economic assessment. *Renewable Energy*, 115, 382–390. https://doi.org/10.1016/j.renene.2017.08.031.

Suhail, R., Chen, J.-F., Gentry, R., et al. (2019). Analysis and design of a pedestrian bridge with decommissioned FRP wind blades and concrete. *Proceedings of the 14th International Symposium on Fiber-Reinforced Polymer Reinforcement of Concrete Structures (FRPRCS)*, Belfast, Northern Ireland, UK.

US Department of Energy. (2023). *Wind energy's economic impacts to communities.* Office of Energy Efficiency and Renewable Energy. https://windexchange.energy.gov/end-of-service-guide.

Vestas. (2023). *Two main blade recycling initiatives.* www.vestas.com/en/sustainability/environment/zero-waste.

Wind Energy Technologies Office. (2021). *Wind repowering helps set the stage for energy transition.* www.energy.gov/eere/wind/articles/wind-repowering-helps-set-stage-energy-transition.

WindEurope. (2020). *Accelerating wind turbine blade circularity.* https://cefic.org/app/uploads/2020/05/Accelarating-wind-turbine-blade-waste-recycling.pdf.

WindEurope. (2022). *Repowering Europe's wind farms is a win-win-win.* https://windeurope.org/newsroom/press-releases/repowering-europes-wind-farms-is-a-win-win-win.

Wiser, R., Bolinger, M., Hoen, B., et al. (2022). *Land-based wind market report: 2022 edition.* US Department of Energy.

Woo, S. M., & Whale, J. (2022). A mini-review of end-of-life management of wind turbines: Current practices and closing the circular economy gap. *Waste Management & Research*, 40(12), 1730–1744. https://doi.org/10.1177/0734242X221105434

Zambon, P., Ricci, P., Bovo, E., et al. (2007). Sarcoma risk and dioxin emissions from incinerators and industrial plants: A population-based case-control study (Italy). *Environmental Health*, 6(1), 19. https://doi.org/10.1186/1476-069X-6-19.

Zhang, L., Yang, J., Chang, Y., et al. (2020). Understanding the material efficiency of the wind power sector in China: A spatial-temporal assessment. *Resources, Conservation & Recycling*, 155, 104668. https://doi.org/10.1016/j.resconrec.2019.104668.

17 Enhancing waste management systems in Canada

Integration of geographic information system and remote sensing approaches

Nima Karimi, Amy Richter and Kelvin Tsun Wai Ng

Introduction

Solid waste management has become increasingly complex in Canada due to rapid urbanization and a growing population (Mohareb et al., 2008; City of Ottawa, 2024). While landfilling remains the primary method of solid waste management in Canada (Vu et al., 2017; Pan et al., 2019) improvements in various aspects of this practice—such as identifying optimal locations for future landfills to minimize environmental impacts and enhance accessibility—can significantly aid in managing the increasing waste footprint.

By leveraging geographical information system (GIS) and remote sensing (RS), this chapter illustrates potential improvements in waste management efficiency, environmental monitoring, and decision-making processes, ultimately contributing to more sustainable waste management practices across Canada. This chapter aims to demonstrate how GIS can be used to identify suitable areas for future landfills, detect adverse effects on neighboring sites using RS indices, identify fugitive gas hotspots in landfills through high-resolution satellite imagery, and regionalize waste management systems by improving landfill coverage and population accessibility. These tools ultimately will enhance the sustainability and efficiency of solid waste management in Canada by optimizing landfill siting, monitoring environmental impacts, and improving accessibility to waste disposal services, contributing to more effective and sustainable waste management practices.

Role of GIS and RS in solid waste management

GIS-based methods for facility siting and prioritizing

The integration of GIS and RS technologies is crucial to effectively site waste disposal facilities. In 2020, over 25 million tonnes of solid waste were disposed of in Canadian landfills, indicating a heavy reliance on this waste management method (Statistics Canada, 2024). In addition, despite the increase in the diversion rate of solid waste in Canada to 27.5% in 2020, the total waste generated rose by 17% from 2002 to 2020, resulting in an 8% increase in waste disposed of in landfills or incinerated, emphasizing the ongoing need for more landfill capacity (Statistics

DOI: 10.4324/9781003534129-23

Canada, 2024). Therefore, careful siting of these facilities is necessary to address environmental impacts and public concerns (Richter et al., 2019). For example, better-located facilities improve accessibility, thereby reducing the occurrence of illegal disposal sites. Additionally, such facilities are more likely to be situated away from water resources, minimizing the risk of water pollution, and are presumably distant from agricultural lands, forest lands, and protected areas, ensuring that these valuable resources remain safeguarded (Karimi et al., 2020, 2021a). Better-located facilities also presumably address the "Not In My Backyard" (NIMBYism) concern, thereby retaining public satisfaction.

Given the size of Canada, GIS and RS technologies can also be more effective in evaluating various spatial and environmental factors essential for waste facility siting. These tools allow the incorporation of multiple data layers, including land use, elevation, and physical characteristics, to conduct thorough spatial analyses. For example, Moeinaddini et al. (2010) and Gorsevski et al. (2012) demonstrated the use of GIS and the Analytic Hierarchy Process (AHP) in landfill siting by integrating diverse data layers for detailed spatial analysis in Malaysia and Macedonia, respectively. Some environmental issues associated with landfills include but are not limited to: leachate generation, toxic gas emissions, and the spread of odors and pollutants like methane and carbon dioxide (Pan et al., 2019; Bruce et al., 2017; Delkash et al., 2016).

Besides GIS and RS, the use of Multi-criteria Decision-Making (MCDM) tools, particularly the AHP, is essential in landfill siting. These tools enable the systematic evaluation of various parameters, facilitating balanced decision-making. AHP, developed by Thomas Saaty, helps break down complex decisions into more manageable subproblems, which can be analyzed independently (Saaty, 1980). Studies have shown the effectiveness of MCDM tools in improving the landfill siting process. For instance, Colvero et al. (2018) enhanced landfill suitability assessments by integrating a fuzzy-AHP approach, which handled uncertainties in criteria weighting. Such methodologies can also be applied in Canada to ensure more reliable site selection, addressing multiple factors including but not limited to environmental, economic, and community concerns.

In addition, by incorporating approaches like fuzzy-AHP, as seen in a study by Colvero et al. (2018), uncertainties in criteria weighting can be managed effectively. Figure 17.1 provides a schematic overview of waste facility siting considering multiple parameters.

Canadian cases, two regional (prairies) and one national studies

Richter et al. (2019) conducted a study to rank locations for waste disposal site expansion around Moose Jaw, Southern Saskatchewan. The researchers integrated and scored various RS indices and vector files. Specifically, they utilized indices such as NDVI (Normalized Difference Vegetation Index), NDBI (Normalized Difference Built-up Index), NDSI (Normalized Difference Snow Index), NDMI (Normalized Difference Moisture Index), and NTL (Night Time Light). Vector files included data on populated places, roads, lakes, rivers, and protected areas. By

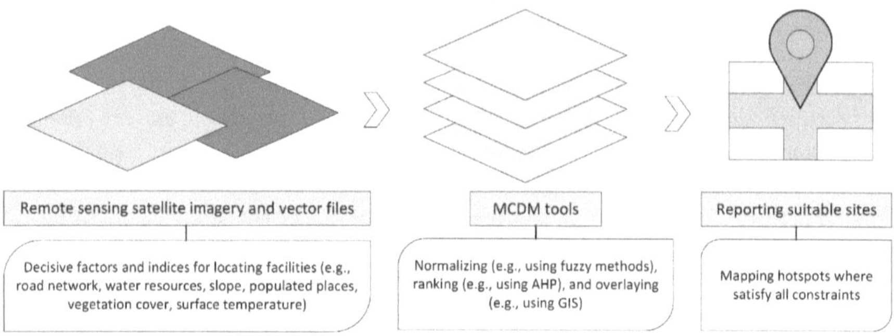

Figure 17.1 Integration of RS and GIS for identifying suitable sites for future waste facilities or ranking the suitability of existing facilities.

integrating these human and environmental factors into a detailed assessment, the study identified locations, for example, western edges of the study area, where sustainable and effective waste disposal expansion can take place. The method provided a comprehensive analysis that balanced environmental protection with practical considerations for waste management.

In the City of Regina, Saskatchewan, Karimi et al. (2020) developed a comprehensive methodology to rank municipal landfill sites by integrating nighttime satellite imagery with the AHP algorithm. The study employed key factors such as height, slope, proximity to water resources, urban areas, protected areas, and land surface temperature to evaluate potential sites. These factors were integrated into a GIS framework to provide a data-driven approach to landfill siting. Karimi et al. (2020) found that road networks and protected areas were the most critical factors, accounting for 46.1% of the overall weight, emphasizing the need for landfills to be both accessible and environmentally responsible. The methodology also involved creating Thiessen polygons based on populated centers defined by NTL imagery, which facilitated the objective assessment of sites without political or administrative boundaries. The resulting suitability map helped identify regions that are far from water resources and protected areas, but in the vicinity of major road networks, making them ideal for new landfill locations (Karimi et al., 2020).

Islam et al. (2023) developed a comprehensive framework for selecting optimal sites for municipal solid waste to value-added (W2VA) facilities across Canada. The framework involved a three-stage decision-making model comprising exclusion analysis, preference analysis, and network analysis. The fuzzy analytic hierarchy process (FAHP) and GIS were integrated to prioritize preference factors and facility locations, resulting in a land suitability map (LSM). Islam et al. (2023) identified 10 optimal sites for W2VA facilities in Western Canada and 15 in Eastern Canada, based on the minimization of transportation distances. The framework considered various social, economic, and environmental factors to ensure sustainable waste management. The adaptable and effective approach highlighted its potential for broader applications in waste management and other environmental planning processes (Islam et al., 2023).

RS indices for facility monitoring and identification

The United Nations' Intergovernmental Panel on Climate Change (IPCC) reports that waste disposal significantly contributes to greenhouse gas emissions, such as carbon dioxide, methane, and nitrogen oxides, accounting for about five percent of total global greenhouse gas generation (IPCC, 2019). Additionally, leachate from anaerobic waste decomposition has been linked to soil and groundwater contamination, impacting human and plant health (Xu et al., 2018). In addition, the effects of land cover changes, due to the presence of waste facilities, on land surface temperature (LST) have been well documented (Ramachandra et al., 2018; Ullah et al., 2019; Yang et al., 2020). For example, the heat generated by waste decomposition at landfill sites can also be significant, affecting the local microclimate and vegetation health (Hao et al., 2017). Elevated LSTs have been reported at various landfill sites globally, highlighting the impact of waste management practices on environmental conditions (Koerner and Koerner, 2006; Hanson et al., 2009; Cai et al., 2014).

On the other hand, Phil-Eze (2010) noted that the disposal and degradation of municipal waste can alter soil chemistry and impact nearby plant species' structure and fertility. Elevated soil metal concentrations near dump sites have been found at significant depths (Azeez et al., 2011), and field studies have confirmed lower metabolism rates and diversity in plant species around dumpsites (Ali et al., 2014).

RS indices are believed to provide an indirect and cost-effective way to assess and delineate polluted areas for disposal sites. For example, elevated LST values, retrieved from thermal bands of Landsat satellite imagery, and lowered NDVI, retrieved from infrared and near-infrared bands of Landsat satellite imagery, around landfills indicate areas of environmental stress, including vegetation degradation and soil contamination (Mahmood et al., 2017, 2019). On the other hand, elevated LST is often associated with increased methane emissions due to the heat generated during waste decomposition (Fjelsted et al., 2019; Javadinejad et al., 2019). By leveraging satellite imagery, these studies used LST to pinpoint methane gas hotspots in landfills, facilitating targeted mitigation efforts and enhancing the accuracy of environmental impact assessments. Figure 17.2 shows how RS indices can be used to monitor and identify waste facility environmental changes.

Canadian case, two national and one local (Toronto) studies

Yan et al. (2014) focused on the Trail Road and Nepean municipal solid waste (MSW) disposal sites in Ottawa, Ontario. They used multi-temporal Landsat satellite images to monitor the LST of these landfills from 1984 to 2011. The study revealed that the average LST of the landfill was consistently higher than the surrounding vegetation and air temperature by 4–10°C and 5–11.5°C, respectively. This difference was more pronounced during the summer. The research highlighted

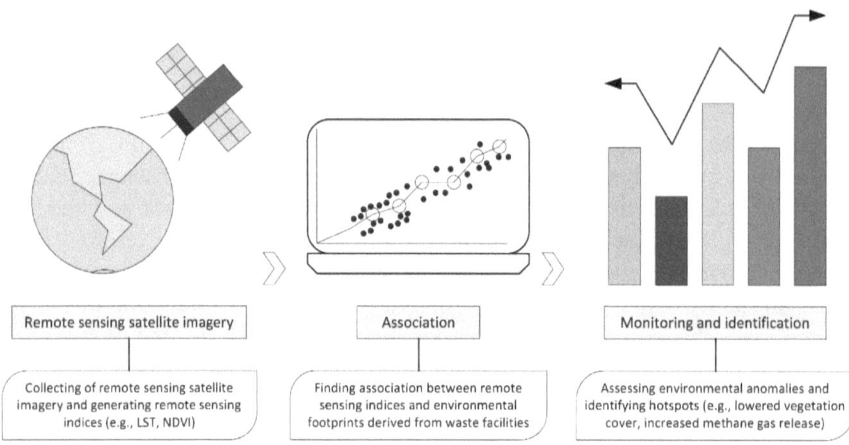

Figure 17.2 Integration of RS imagery in monitoring waste facilities.

that there was no significant temperature difference between the closed Nepean landfill and the active Trail Road landfill until 2007, after which the Trail Road landfill showed a higher LST due to ongoing construction and expansion activities (Yan et al., 2014).

Karimi et al. (2021a) analyzed thermal heterogeneity in the proximity of MSW landfills across eight landfill sites in Canada, selected from multiple provinces including Ontario, Saskatchewan, and British Columbia. The study utilized Landsat-8 and Sentinel-2 imagery from 2015 to 2020 to monitor the landfills. Indices such as LST and NDVI were employed to identify and evaluate biothermal zones within a specific radius around the landfills. Karimi et al. (2021a) found that both the magnitude and variability of LST were higher in agricultural lands compared to forest lands, suggesting that agricultural lands are more sensitive to microclimatic changes caused by landfills. The study revealed that biothermal zones typically extended up to 270 m from the landfill boundaries. Elevated LST within these zones indicated increased environmental stress, while lowered NDVI values suggested vegetation degradation. It is believed that with the extent and characteristics of biothermal zones, policymakers and landfill operators can develop targeted strategies to monitor and reduce environmental footprints and protect surrounding ecosystems.

Karimi et al. (2021b) developed an approach to identify methane emission hotspots using Sentinel-2 imagery and Landsat satellite imagery for the Green Lane landfill in Toronto, Ontario. Sentinel-2 was utilized for its high-resolution bands, while Landsat was employed for its thermal band capabilities. This combination allowed for the generation of thermal imagery with a 10-m resolution once the training phase for satellite imagery integration was completed using the Random Forest (RF) method. The satellite images were retrieved on June 15, 2020, and July

20, 2020, chosen to capture seasonal variations in landfill emissions. The RF model was trained with a dataset that included thermal and spectral bands from Landsat and Sentinel-2 respectively. The accuracy of the model was validated using ground truth data, achieving a high accuracy rate of 92%. The results estimated several methane emission hotspots within the landfill. The thermal imagery generated by the RF model provided detailed insights into the spatial distribution of these hotspots, allowing for more precise monitoring and management of landfill emissions (e.g., locating methane gas extraction wells).

GIS network analysis and RS imagery for regionalizing waste facilities

By handling extensive geospatial data and multiple spatial variables, GIS network analysis improves the efficiency of waste collection and transportation. For instance, it can manage the strategic placement of storage bins, demand points, waste facilities, road types, and speed limits, which leads to more efficient routing and reduced travel times and distances (Xue et al., 2015). Additionally, GIS tools can incorporate Vehicle Routing Problem (VRP) algorithms to further refine collection systems by evaluating practical scenarios, such as adjusting travel schedules and considering capacity constraints for collection trucks (Huang & Lin, 2015; Vu et al., 2018, 2020).

Considering that a significant portion of waste management budgets is devoted to collection costs—ranging from 40% to 80% in many countries (Kinobe et al., 2015)—accurately identifying waste generation points and strategically designing landfill locations and collection schedules remain critical. In addition, effective waste management systems should ensure that the majority of population centers have access to waste collection services. The lack of well-placed landfills and efficient collection systems can lead to illegal dumping and environmental hazards (Glanville & Chang, 2015; Mahmood et al., 2017; Richter et al., 2020).

Integration of GIS network analysis with satellite imagery is specifically important in regions where population centers are not well defined, such as rural and remote areas. In addition, RS imagery, such as NTL, has been used to identify anthropogenic activities (e.g., recovery after natural disasters) and population points, supporting more informed decision-making and service delivery (Wang et al., 2018; Enenkel et al., 2020).

In Canada, the application of these technologies is especially crucial because of remote northern communities and their limited access to waste facilities. Many small towns and temporary settlements in rural Canada lack precise mapping and reliable records of their size and status (Karimi et al., 2022a). Dispersed populations, remote northern communities, and their inaccessibility to proper solid waste management systems raise significant issues. For example, in some First Nation communities, these challenges can lead to improper waste disposal practices, such as open burning and illegal dumping, which consequently pose significant health and environmental risks (Bharadwaj et al., 2006; Keske et al., 2018). In addition, six First Nation communities in Alberta and Saskatchewan have reported illegal disposal sites as major threats to their drinking water resources (Patrick, 2018).

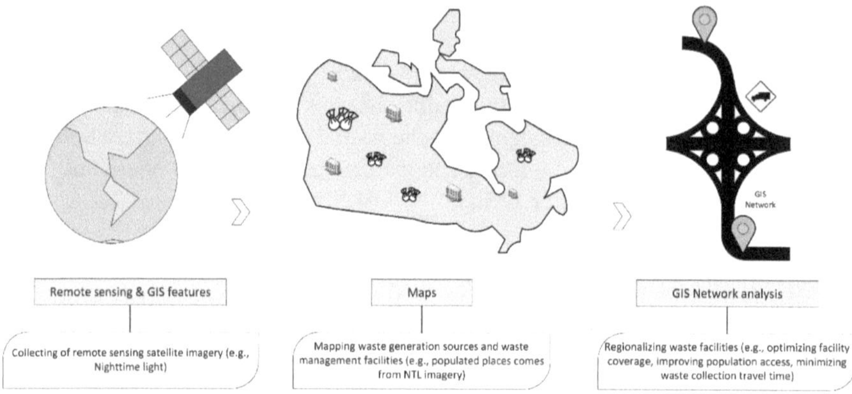

Figure 17.3 Integration of RS and GIS network analysis in regionalizing waste management.

Adopting a regionalized approach to waste management can help address these challenges. By accurately identifying population centers, including those in remote areas, waste management systems can strategically position landfills to minimize travel distances for collection trucks and improve coverage. This approach enhances both accessibility of waste services and the overall efficiency of waste collection operations, reducing operational costs and improving services delivery in underserved regions. Figure 17.3 schematically shows the steps considered for the adoption of GIS and RS in waste facility regionalization.

Canadian cases, two national and one regional (Saskatchewan) studies

Richter, Ng, Karimi et al. (2019) optimized waste management regions in Canada using GIS and ArcGIS Model Builder, aiming to enhance efficiency and reduce operational costs by minimizing the standard deviation of landfills, populated places, and road lengths. Data from Canadian federal and municipal sources, including federal census subdivisions and historical waste management regions for Nova Scotia and transportation planning committee regions for Saskatchewan, were utilized. Recursive Thiessen polygons created initial tessellations and were refined iteratively. The results indicated improved spatial distribution of waste management facilities, suggesting that tailored optimization strategies can significantly enhance the efficiency of waste management systems. These findings have practical implications for regional waste management planning, highlighting the need for data-driven approaches.

Karimi et al. (2022a) optimized population coverage by landfills using network analysis and satellite imagery. Utilizing NASA Black Marble NTL imagery to identify population centers, GIS network analysis tools were employed to maximize landfill coverage and minimize travel distances for waste collection trucks. Data was collected from three Canadian provinces—Ontario, Saskatchewan, and British Columbia—covering rural areas. Their results demonstrated that the proposed data-driven analytical framework improved landfill regionalization and waste collection efficiency. Performance indicators showed better landfill regionalization index (LFRI) scores, reduced travel times, and enhanced landfill area coverage compared to the status quo. These findings underscore the effectiveness of integrating RS data and GIS network analysis to develop sustainable and efficient waste management systems at a regional scale.

Similarly, Karimi et al. (2022b) developed and evaluated two regionalized optimization approaches using GIS network analysis tools, addressing both MSW collection and landfill siting across 18 Saskatchewan census divisions. They also used NASA Black Marble satellite imagery to identify population centers, the study provided two novel performance indicators as "Accessibility" and "Landfill Minimization" indices. Their findings reveal improvements in landfill accessibility and efficiency, particularly in border-sharing divisions. Furthermore, they highlighted the importance of an established road network for effective waste management and suggested that regional approaches can improve service delivery, particularly for sparsely populated areas and indigenous communities.

Discussion

The integration of GIS and RS technologies in waste management offers numerous advantages. Their ability to handle large datasets improves the efficiency and accuracy of waste management planning in urban/rural arial and remote communities. However, it is also believed that the implementation of these tools requires specialized training, which can be time-consuming for governmental bodies or managers responsible for their incorporation. In addition, the lack of awareness and familiarity among decisionmakers and practitioners, reliance on traditional methods due to regulatory agencies and consultants' resistance to change, and the costs of the initial investment in technology and training particularly for organizations with limited budgets might limit these tools' widespread use.

Looking ahead, the future of waste management will likely see greater integration of GIS and RS technologies, as evidenced by case studies in different regions of Canada in this chapter. These technologies can further enhance the accuracy and efficiency of waste management systems, enabling real-time monitoring and predictive modeling. However, this could only be possible by concerted efforts in capacity building, awareness campaigns, and policy reforms to promote the adoption of these innovative technologies.

Conclusion

This chapter has highlighted the role of GIS and RS technologies in improving waste management systems in Canada. Examples include the GIS-based methods for landfill siting and expansion, demonstrated in Regina and Moose Jaw, Saskatchewan, to effectively optimize site selection by considering various factors. RS indices help monitor landfill impacts, such as identifying fugitive gas hotspots and temperature anomalies, and providing cost-effective and accurate environmental assessments. The national study spanning Eastern to Western Canada shows the potential of combining GIS network analysis and satellite imagery to improve landfill coverage and accessibility. It is believed that addressing challenges such as cost, complexity, and institutional barriers can help with GIS and RS widespread adoption by relevant stakeholders including municipal governments, waste management agencies, and environmental planners.

References

Ali, S. M., Pervaiz, A., Afzal, B., Hamid, N., & Yasmin, A. (2014). Open dumping of municipal solid waste and its hazardous impacts on soil and vegetation diversity at waste dumping sites of Islamabad city. *Journal of King Saud University-Science*, 26, 59–65 https://doi.org/10.1016/j.jksus.2013.08.003.

Azeez, J. O., Hassan, O. A., & Egunjobi, P. O. (2011). Soil contamination at dumpsites: Implication of soil heavy metals distribution in municipal solid waste disposal system: A case study of Abeokuta, southwestern Nigeria. *Soil and Sediment Contamination: An International Journal*, 20, 370–386 https://doi.org/10.1080/15320383.2011.571312.

Bharadwaj, L., Nilson, S., Judd-Henrey, I., Ouellette, G., Parenteau, L., Tournier, C., Watson, D., Bear, D., Ledoux, G., & Bear, A. (2006). Waste disposal in First-Nations communities: The issues and steps toward the future. *Journal of Environmental Health*, 68(7), 35–39. Accessed on https://search.ebscohost.com/login.aspx?direct=true&db=a9h&AN=20052056&site=ehost-live by 2021/Nov/25.

Bruce, N., Ng, K. T. W., & Richter, A. (2017). Alternative carbon dioxide modeling approaches accounting for high residual gases in LandGEM. *Environmental Science and Pollution Research*, 24(16), 14322–14336. https://dx.doi.org/10.1007/s11356-017-8990-9.

Cai, B.-F., Jian-Guo, L., Qing-Xian, G., Xiao-Qin, N., Dong, C., Lan-Cui, L., Ying, Z., & Zhan-Sheng, Z. (2014). Estimation of methane emissions from municipal solid waste landfills in China based on point emission sources. *Advances in Climate Change Research*, 5, 81–91. https://doi.org/10.3724/SP.J.1248.2014.081.

City of Ottawa. (2024). *Solid waste mater plan*. Accessed on https://engage.ottawa.ca/12201/widgets/183278/documents/132245 by 2024/Jul/13.

Colvero, D. A., Gomes, A. P. D., Tarelho, L. A. d. C., Matos, M. A. A. d., & Santos, K. A. d. (2018). Use of a geographic information system to find areas for locating of municipal solid waste management facilities. *Waste Management,* 77, 500–515. https://doi.org/10.1016/j.wasman.2018.04.036.

Delkash, M., Zhou, B., Han, B., Chow, F. K., Rella, C. W., & Imhoff, P. T. (2016). Short-term landfill methane emissions dependency on wind. *Waste Management*, 55, 288–298 https://doi.org/10.1016/j.wasman.2016.02.009.

Enenkel, M., Shrestha, R. M., Stokes, E., Román, M., Wang, Z., Espinosa, M. T. M., Hajzmanova, I., Ginnetti, J., & Vinck, P. (2020). Emergencies do not stop at night: Advanced analysis of displacement based on satellite-derived nighttime light observations. *IBM Journal of Research and Development*, 64, 8:1–8:12. https://doi.org/10.1147/JRD.2019.2954404.

Fjelsted, L., Christensen, A., Larsen, J., Kjeldsen, P., & Scheutz, C. (2019). Assessment of a landfill methane emission screening method using an unmanned aerial vehicle mounted thermal infrared camera: A field study. *Waste Management*, 87, 893–904. https://doi.org/10.1016/j.wasman.2018.05.031.

Glanville, K., & Chang, H. (2015). Mapping illegal domestic waste disposal potential to support waste management efforts in Queensland, Australia. *International Journal of Geographical Information Science*, 29(6), 1042–1058. https://doi.org/10.1080/13658816.2015.1008002.

Gorsevski, P. V., Donevska, K. R., Mitrovski, C. D., & Frizado, J. P. (2012). Integrating multi-criteria evaluation techniques with geographic information systems for landfill site selection: A case study using ordered weighted average. *Waste Management*, 32, 287–296. https://doi.org/10.1016/j.wasman.2011.09.023.

Hanson, J. L., Yeşiller, N., & Oettle, N. K. (2009). Spatial and temporal temperature distributions in municipal solid waste landfills. *Journal of Environmental Engineering*, 136, 804–814. https://doi.org/10.1061/(ASCE)EE.1943-7870.0000202.

Hao, Z., Sun, M., Ducoste, J., Benson, C., Luettich, S., Castaldi, M., & Barlaz, M. (2017). Heat generation and accumulation in municipal solid waste landfills. *Environmental Science & Technology*, 21, 12434–12442. https://pubs.acs.org/doi/abs/10.1021/acs.est.7b01844.

Huang, S., & Lin, P. (2015). Vehicle routing–scheduling for municipal waste collection system under the "Keep Trash off the Ground" policy. *Omega*, 55, 24–37. https://doi.org/10.1016/j.omega.2015.02.004

IPCC. (2019). *Waste management*. Retrieved from www.ipcc.ch/site/assets/uploads/2018/02/ar4-wg3-chapter10-1.pdf on 2019/Oct/24.

Islam, M. S., Sebastian, R. M., Kurian, V., Billal, M. M., & Kumar, A. (2023). The development of a framework for the selection of optimal sites for the location of municipal solid waste to value-added facilities through the integration of a geographical information system and a fuzzy analytic hierarchy process. *Biofuels, Bioproducts and Biorefining*, 17(6), 1622–1638. https://doi.org/10.1002/bbb.2530.

Javadinejad, S., Eslamian, S., & Ostad-Ali-Askari, K. (2019). Investigation of monthly and seasonal changes of methane gas with respect to climate change using satellite data. *Applied Water Science,* 9, 180 https://doi.org/10.1007/s13201-019-1067-9.

Karimi, N., Ng, K. T. W., & Richter, A. (2022a). Integrating geographic information system network analysis and nighttime light satellite imagery to optimize landfill regionalization on a regional level. *Environmental Science and Pollution Research*, 29(54), 81492–81504. https://doi.org/10.1007/s11356-022-21462-w.

Karimi, N., Ng, K. T. W., & Richter, A. (2022b). Development of a regional solid waste management framework and its application to a prairie province in central Canada. *Sustainable Cities and Society*, 82, 103904. https://doi.org/10.1016/j.scs.2022.103904.

Karimi, N., Ng, K. T. W., Richter, A., Williams, J., & Ibrahim, H. (2021a). Thermal heterogeneity in the proximity of municipal solid waste landfills on forest and agricultural lands. *Journal of Environmental Management*, 287, 112320. https://doi.org/10.1016/j.jenvman.2021.112320.

Karimi, N., Ng, K. T. W., & Richter, A. (2021b). Prediction of fugitive landfill gas hotspots using a random forest algorithm and Sentinel-2 data. *Sustainable Cities and Society*, 73, 103097. https://doi.org/10.1016/j.scs.2021.103097.

Karimi, N., Richter, A., & Ng, K. T. W. (2020). Siting and ranking municipal landfill sites in regional scale using nighttime satellite imagery. *Journal of Environmental Management*, 256, 109942. https://doi.org/10.1016/j.jenvman.2019.109942.

Keske, C. M., Mills, M., Godfrey, T., Tanguay, L., & Dicker, J. (2018). Waste management in remote rural communities across the Canadian north: Challenges and opportunities. *Detritus*, 2(1), 63. https://doi.org/10.31025/2611-4135/2018.13641.

Kinobe, J. R., Bosona, T., Gebresenbet, G., Niwagaba, C. B., & Vinnerås, B. (2015). Optimization of waste collection and disposal in Kampala city. *Habitat International*, 49, 126–137. https://doi.org/10.1016/j.habitatint.2015.05.025.

Koerner, G. R., & Koerner, R. M. (2006). Long-term temperature monitoring of geomembranes at dry and wet landfills. *Geotextiles and Geomembranes*, 24(1), 72–77. https://doi.org/10.1016/j.geotexmem.2004.11.003.

Mahmood, K., Batool, A., Faizi, F., Chaudhry, M. N., Ul-Haq, Z., Rana, A. D., & Tariq, S. (2017). Bio-thermal effects of open dumps on surroundings detected by remote sensing: Influence of geographical conditions. *Ecological Indicators*, 82, 131–142. https://doi.org/10.1016/j.ecolind.2017.06.042.

Mahmood, K., Ul-Haq, Z., Faizi, F., & Batol, S. A. (2019). A comparison of satellite-based indices for hazard assessment of MSW open dumps using spatial analysis. *Waste Management & Research*, 37, 219–236. https://doi.org/10.1177/0734242X18815963.

Moeinaddini, M., Khorasani, N., Danehkar, A., Darvishsefat, A. A., & Zienalyan, M. (2010). Siting MSW landfill using weighted linear combination and analytical hierarchy process (AHP) methodology in GIS environment (case study: Karaj). *Waste Management*, 30, 912–920. https://doi.org/10.1016/j.wasman.2010.01.015.

Mohareb, A. K., Warith, M. A., & Diaz, R. (2008). Modelling greenhouse gas emissions for municipal solid waste management strategies in Ottawa, Ontario, Canada. *Resources, Conservation and Recycling*, 52(11), 1241–1251. https://doi.org/10.1016/j.rescon rec.2008.06.006.

Pan, C., Ng, K. T. W., and Richter, A. (2019). An integrated multivariate statistical approach for the evaluation of spatial variations in groundwater quality near an unlined landfill. *Environmental Science and Pollution Research*, 26(6), 5724–5737. https://doi.org/10.1007/s11356-018-3967-x.

Patrick, R. J. (2018). Adapting to climate change through source water protection: Case studies from Alberta and Saskatchewan, Canada. *International Indigenous Policy Journal*, 9(3). http://dx.doi.org/10.18584/iipj.2018.9.3.1.

Phil-Eze, P. O. (2010). Variability of soil properties related to vegetation cover in a tropical rainforest landscape. *Journal of Geography and Regional Planning*, 3(7), 177–184.

Ramachandra, T. V., Bharath, S., & Gupta, N. (2018). Modelling landscape dynamics with LST in protected areas of Western Ghats, Karnataka. *Journal of Environmental Management*, 206, 1253–1262. https://doi.org/10.1016/j.jenvman.2017.08.001.

Richter, A., Ng, K. T. W., & Karimi, N. (2019). A data driven technique applying GIS, and remote sensing to rank locations for waste disposal site expansion. *Resources, Conservation and Recycling,* 149, 352–362. https://doi.org/10.1016/j.resconrec.2019.06.013.

Richter, A., Ng, K. T. W., Karimi, N., & Chang, W. (2020). Developing a novel proximity analysis approach for assessment of waste management cost efficiency in low population density regions. *Sustainable Cities and Society*, 102583. https://doi.org/10.1016/j.scs.2020.102583.

Richter, A., Ng, K. T. W., Karimi, N., Wu, P., & Kashani, A. H. (2019). Optimization of waste management regions using recursive Thiessen polygons. *Journal of Cleaner Production*, 234, 85–96. https://doi.org/10.1016/j.jclepro.2019.06.178.

Saaty, T.L. (1980). *The analytical hierarchy process, planning, priority. Resource allocation.* RWS Publications.

Statistics Canada. (2024). *Solid waste diversion and disposal, environmental indicators*. Accessed on www.canada.ca/en/environment-climate-change/services/environmental-indicators/solid-waste-diversion-disposal.html by 2024/Jul/17.

Ullah, S., Ahmad, K., Sajjad, R.U., Abbasi, A.M., Nazeer, A., & Tahir, A.A. (2019). Analysis and simulation of land cover changes and their impacts on land surface temperature in a lower Himalayan region. *Journal of Environmental Management*, 245, 348–357. https://doi.org/10.1016/j.jenvman.2019.05.063.

Vu, H. L., Ng, K. T. W., & Bolingbroke, D. (2018). Parameter interrelationships in a dual phase GIS-based municipal solid waste collection model. *Waste Management*, 78, 258–270 https://doi.org/10.1016/j.wasman.2018.05.050.

Vu, H. L., Ng, K. T. W., Fallah, B., Richter, A., & Kabir, G. (2020). Interactions of residential waste composition and collection truck compartment design on GIS route optimization. *Waste Management*, 102, 613–623. https://doi.org/10.1016/j.wasman.2019.11.028.

Vu, H. L., Ng, K. T. W., Richter, A. (2017). Optimization of first order decay gas generation model parameters for landfills located in cold semi-arid climates. *Waste Management*, 69, 315–324. https://dx.doi.org/10.1016/j.wasman.2017.08.028

Wang, Z., Román, M. O., Sun, Q., Molthan, A. L., Schultz, L. A., & Kalb, V. L. (2018c). Monitoring disaster-related power outages using NASA black marble nighttime light product. *International Archives of the Photogrammetry, Remote Sensing and Spatial Information Sciences*, XLII-3, 1853–1856. https://doi.org/10.5194/isprs-archives-XLII-3-1853-2018.

Xu, Y., Xue, X., Dong, L., Nai, C., Liu, Y., & Huang, Q. (2018). Long-term dynamics of leachate production, leakage from hazardous waste landfill sites and the impact on groundwater quality and human health. *Waste Management*, 82, 156–166 https://doi.org/10.1016/j.wasman.2018.10.009.

Xue, W., Cao, K., & Li, W. (2015). Municipal solid waste collection optimization in Singapore. *Applied Geography*, 62, 182–190. https://doi.org/10.1016/j.apgeog.2015.04.002.

Yan, W. Y., Mahendrarajah, P., Shaker, A., Faisal, K., Luong, R., & Al-Ahmad, M. (2014). Analysis of multi-temporal Landsat satellite images for monitoring land surface temperature of municipal solid waste disposal sites. *Environmental Monitoring and Assessment*, 186(12), 8161–8173. https://doi.org/10.1007/s10661-014-3995-z.

Yang, C., Yan, F., & Zhang, S. (2020). Comparison of land surface and air temperatures for quantifying summer and winter urban heat island in a snow climate city. *Journal of Environmental Management*, 265, 110563. https://doi.org/10.1016/j.jenvman.2020.110563.

Index

Note: Figures are indicated by *italics*. Tables are indicated by **bold**.

Kolibri 360 GmbH
Ferdinand-Jühlke-Straße 7
99095 Erfurt, Deutschland
produktsicherheit@kolibri360.de